The Ravenous Brain

The Ravenous Brain

How the New Science of Consciousness
Explains Our Insatiable Search for Meaning

Daniel Bor

BASIC BOOKS

A Member of the Perseus Books Group
New York

Published by Basic Books,

A Member of the Perseus Books Group

Books published by Basic Books are available at special discounts for bulk purchases in the United States by corporations, institutions, and other organizations. For more information, please contact the Special Markets Department at the Perseus Books Group, 2300 Chestnut Street, Suite 200, Philadelphia, PA 19103, or call (800) 810-4145, ext. 5000, or e-mail special.markets@perseusbooks.com.

Library of Congress Cataloging-in-Publication Data
Bor, Daniel.
    The ravenous brain : how the new science of consciousness explains our insatiable search for meaning / Daniel Bor.
        p. cm.
    Includes bibliographical references and index.
    ISBN 978-0-465-02047-8 (hardcover : alk. paper) — ISBN 978-0-465-03296-9 (e-book)
1. Consciousness—Physiological aspects. 2. Brain. 3. Mind and body. I. Title.

QP411.B59 2012
612.8'2—dc23

                                                                        2012016971

10 9 8 7 6 5 4 3 2 1

Dedicated to the memory of my father,
Rayle Jonathan Bor, with much love

Contents

Introduction   ix

1   Conceptual Conundrums of Consciousness:
Philosophy   1

2   A Brief History of the Brain:
Evolution and the Science of Thought   35

3   The Tip of the Iceberg:
Unconscious Limits   79

4   Pay Attention to That Pattern!
Conscious Contents   109

5   The Brain's Experience of a Rose:
Neuroscience of Awareness   157

6   Being Bird-Brained Is Not an Insult:
Uncovering Alien Consciousness   195

7   Living on the Fragile Edge of Awareness:
Profound Brain Damage and Disorders of Consciousness   221

8   Consciousness Squeezed, Stretched, and Shrunk:
Mental Illness as Abnormal Awareness   235

Epilogue:
A Delicious Life   265

Acknowledgments   273
Notes and References   275
Illustration Credits   307
Index   309

Introduction

On the night of Thursday, May 8, 1997, my father had a stroke. I'd been repeatedly reassured that the stroke was "very minor." Nevertheless, when I visited him in the hospital, I felt profoundly disturbed by what I witnessed. This sluggish, exhausted man in front of me looked like my father, but I knew, deep down, that he wasn't.

There were subtle clues that betrayed this impostor. Some changes verged on the comical, such as his newfound obsession for Kit Kats—he would eat nothing else for days. Others were more disconcerting. These differences could generally be characterized as a reversion from a sharp, responsible man into a confused child. Even more bizarrely, his attitude toward me would radically alter depending on whether I sat on the right or left side of his bed. When I sat on his right, he would take an interest in me, and we'd have a semi-coherent conversation. When I went instead to his left, it was as though I wasn't in the room. He simply wasn't aware of my presence.

I found myself morbidly wishing that he'd suffered a mild heart attack instead of a stroke. Then, at least, my dad would still be alive, as my dad. As it was, if this situation persisted, a portion of my father would already have died, and every time he spoke, I would be reminded of that fragment of his identity that was lost.

Amid my sense of shock at this new person that wasn't Dad, and my chest-gripping anxiety that he would never fully recover, I couldn't help examining his symptoms dispassionately. My father's stroke had struck a few weeks before my university finals. I was studying philosophy and biological psychology, and consciousness was a hot topic in both fields. On the one hand, I was revising elegant philosophical arguments proposing that consciousness was nonphysical and had little to do with brains. On the other hand, I was poring over the evidence for whether consciousness lay in this cortical region or that, and learning the details of "neglect"—the common stroke condition my father showed by ignoring the left side of space.

Sitting by my father's bedside, I felt sure that the esoteric philosophical position was alien—so mistaken as almost to be offensive. Here was a man I loved dearly, robbed of his identity because a small clot on his brain had potently wounded his consciousness. *Of course consciousness is a physical thing*, I thought, as I sat on his left, achieving the painful magic trick of turning invisible. I didn't exactly know why the philosophical arguments were flawed, or which brain theory of consciousness was most compelling at the time, but I did know which road I wanted to take to find out.

Although previously I was considering a PhD in the philosophy of mind, now there was no contest—a PhD in the neuroscience of consciousness it was. Soon afterward, I was accepted to study this at the University of Cambridge. I've been investigating this and related fields ever since, mainly at Cambridge, but also recently at the newly opened Sackler Centre for Consciousness Science at the University of Sussex. From the first painful glimpse of my father's fractured consciousness, I understood how vital and fundamental this field is, but over the years I've increasingly discovered its fascinating and far-reaching twists and turns. Now I want to share each of these facets with you.

There is nothing more important to us than our own awareness.* We see the breathtaking beauty of snowcapped mountains, the exhilarating grace and speed of a cheetah on a hunt. We hear melodic birdsong in our gardens. We fall in love, or experience the joy of our child's first smile. We compose and appreciate music, art, and literature. We talk and laugh with our friends and family. All these, and everything else we care about, are conscious events. If none of these events were conscious, if we weren't conscious to experience any of them, we'd hardly consider ourselves alive—at least not in any way that matters.

---

* Throughout this book, I will assume that "awareness" and "consciousness" have the same meaning.

When I'm reveling in a glowing pleasure, or even if I'm enduring a sharp sadness, I always sense that behind everything there is the privilege and passion of experience. Our consciousness is the essence of who we perceive ourselves to be. It is the citadel for our senses, the melting pot of thoughts, the welcoming home for every emotion that pricks or placates us. For us, consciousness simply is the currency of life.

However, the scientific study of consciousness, for most of its existence, claimed the prize for the most vital, intimate, meaningful topic with the smallest research interest. Never before have we come within touching distance of understanding the history of our universe, its shape and form, the laws that govern every sparkling star and every dancing atom. Never before have we realized that within every cell in our bodies there lies, coiled up, the code that both defines us and connects us with all life on this planet. All of our wonderful toys of technology and all of our shining scientific discoveries have conscious endeavor to thank for their existence. Yet, until only a couple of decades ago, virtually no one was interested in the science of consciousness, and very little was understood about how the brain generates our experiences.

Historically, it was not scientists who grappled with the conundrum of consciousness, but philosophers. Nearly four hundred years ago, Descartes asserted that consciousness was an entirely personal, subjective entity, impenetrable both to the physical sciences and the minds of other people. When I listen to a Beethoven piano sonata, the sounds I hear, the way that the notes move me, is something that I can only ever imperfectly communicate via the crude medium of words. No one else can ever truly know what I experience—at least that's what many people assume. The peculiarity and power of this observation is highlighted when consciousness is contrasted with any physical entity. Take any object one would care to name, from a subatomic particle to a brick to a star: Thousands of people all could, in principle, explore the same object from different angles, yet uncover an identical set of facts about it. For consciousness, it seems that there is no objectivity. Nor are there multiple viewpoints: There is only one viewpoint—mine.

Modern philosophers have expanded on this foundation, providing arguments for the position that consciousness is elevated beyond the pedestrian whirrings of our brains. Similarly, they claim that the sumptuous, varied menu of feelings and knowledge we experience simply cannot be reduced to some tawdry computer or machine. I begin this book by addressing these stances, rooted in history. There's no denying their intuitive and emotional appeal, but this should always be trumped by the picture that the empirical evidence paints. Indeed, when the light of science shines with forensic detail on this set of philosophical positions, their seeming validity dissolves. Instead, I argue, the most plausible view is that consciousness is a product of the brain, which is a form of computer.

Placing consciousness within the framework of a computational brain suggests a connection between awareness and information processing, since data analysis is the overriding purpose of our inner neuronal world. This general context also implies that our capacity to experience might have an evolutionary heritage, just as our neural machinery does.

Indeed, the common waters between consciousness, information, and biology run deep. A fundamental feature of nature is its ability to store and manipulate information. Evolution ensures that every life-form is a master at hoarding useful "ideas" about the world—not consciously, but via the blind representation of information in its chemical makeup. Consequently, there are countless examples to illustrate that animals aren't the only clever organisms around. Plants have spikes, poisons, and a plethora of other ingenious tools to ward off predators. Even bacteria have an incredibly sophisticated arsenal of weapons designed to infiltrate a host, or thwart a potential assailant. These strategies are written in the language of deoxyribonucleic acid (DNA), a recipe of implicit beliefs about how best to function in the world. But how does this DNA-based system "learn" to build and adapt such accurate blind concepts in the first place?

In any form of learning, from bacteria becoming resistant to antibiotics, to a child making the tentative transition from crawling to walking, there is a tension between holding on to the familiar—to an existing bank of beliefs, a known way of life—and moving toward something new. This novelty injects a bit of chaos into the mix to shake up existing, stubborn ideas. These stumbling movements can potentially make matters worse, but, crucially, they at least allow for the possibility of improvement.

DNA is a fantastic medium for maintaining those stable implicit concepts about the world, and undoubtedly this is why it became the universal carrier for the recipe of life. But there are also mechanisms that can re-arrange the letters of the DNA recipe, so that a new collection of ideas can be written in future generations. When the world has moved on without you, when you and your current DNA-based beliefs are heading toward ex-tinction, new concepts become essential, and this random mixing of the DNA pot might just create a few surviving organisms amid the many that fail. Such winners will in their DNA be carrying successful novel insights to steer them through these difficult times. In other words, this section of the species has blindly learned, over the generations, to innovate its way around a dangerous obstacle.

Without this ability to innovate, albeit in a random, cruel, inefficient manner, life simply would not have persisted on earth, and so this skill of nature to track the changing tides, and to exploit any advantage it can latch onto, is in some ways the essence of evolution. It's no wonder, therefore, that such a fundamental attribute of life would burst the seams of its original DNA dam and spill into new territory. And the premier example of this ex-pansion of learning dimensions is the evolutionary invention of a bundle of specialist computational cells that constitute a brain. Now an extra, panoramic range of innovations can occur within a single organism in a pointed, purposive way as it mentally probes the environment and stores within its neurons any new information, not coded in its DNA, that is rel-evant for survival.

Humans, too, are clearly cast in evolution's intense furnace of the fight for survival. Although our mental life sometimes appears opaque by its sheer complexity, evolution has carefully finessed the foundations of every facet of our biological makeup—including our most sophisticated emotions and most inspired ideas.

Although humans are only one tiny strand of the web of life, we have a unique place in nature because of the vigor of our intellect and the extent of our awareness. We can muster up only about a fifth of the physical strength of our nearest relatives, the chimpanzees; even the sharpness of our senses is feeble compared to our chimp cousins. And yet, along with our supreme consciousness, we in some ways encapsulate evolution's fundamental driving force by being absolute masters of innovation. Every species pushes to con-trol and dominate its environment. But we, via our own ingenuity, have re-shaped a staggering portion of the globe for our own benefit.

I will propose that innovations, those brightest of information-processing gifts, are the main purpose of consciousness. But it's certainly not the case that all flavors of neural information will reach awareness. Many basic computational functions, such as the control of our breathing, can tick along perfectly well without any input from our conscious minds. Widespread, though simple, statistical learning constantly occurs without the aid of our awareness. Also, importantly, if we've previously consciously mastered some skill, such as walking, then our unconscious minds can almost entirely take over such tasks. In all these cases, the staid, pedestrian parts of information processing are handled by our unconsciousness, while our conscious minds are freed to dwell on newer, more difficult topics. And for any lesson involving even a smidgen of novelty or complexity, we simply have to engage our conscious minds to learn it.

Consciousness is the shining, gold-plated experimental laboratory of our mental mansion where we can analyze virtually anything to great depths. So as not to waste time and energy-intensive neural resources, our brains have to be extremely picky about what they let into this prized place. Attention is the gatekeeper of our awareness, only pushing through those items from our senses or inner cogitations that have the most pressing biological salience, and especially those unexpected features that offer us the greatest potential insights.

The space inside this playground of experiences, though, is frustratingly cramped. Our consciousness can only simultaneously deal with about four items, fully processed. But the magic arises by the amazing variety of ways that we can manipulate what resides within our conscious minds, and it is in these shuffling mental actions that we can learn profound truths about the world.

Crucially, our rich experiential landscape reflects the fact that awareness isn't concerned with just any raw snippet of data. In fact, the absolute opposite is the case, as illustrated by the following curious, niggling conundrum: When most animals have fed well and found a safe place to stay, they normally then make the eminently sensible decision to rest. Admittedly, some species occasionally play in ways that look perfectly like practice for hunting or fighting. But that's about it. Humans, in striking contrast, when all biological needs are met, reach for a sudoku puzzle or a games console—or we may even peruse a science book. Astoundingly, a few of us see the extended crossword puzzle of scientific research as a wonderful hobby. We therefore get the biologically perverse situation of Albert Einstein converting the fiendish cryptic clue of the whole universe into the neat five-character solution of $E = mc^2$—in his spare time!

Thus, one defining characteristic of humanity is its ravenous appetite for facts. But we don't hunger for any old ragbag of information—no, we especially crave that small subset of knowledge that involves patterns. Uncovering the hidden structure in a puzzle may seem like a trivial idiosyncrasy of the human mind, as far removed from our evolutionary drive for survival as it is possible to be. But looks can be deceptive. This chronic mental hunger might dabble occasionally in intellectual play, rather like a thoroughbred horse that spontaneously chooses to gallop, just to exercise its sinewy, muscular frame. Much of the time, though, our restless, roving curiosity will latch onto real wisdom—not the limited, quiet wisdom of an old man imparting measured advice, but, in a broader meaning of the term, any bold innovation that suddenly empowers us with impressive new tools of understanding and control. Consciousness as a ravenous appetite for wisdom led us to discover fire, farming, and, indeed, all the modern products of science and technology that make our lives easier, longer, and more entertaining. On a smaller scale, this hunger for innovation, burning with a particularly bright restlessness when we're young, guides us toward our first words and then toward the myriad stepping stones of knowledge needed to negotiate adulthood in a complex, modern world.

Consciousness is that choice mental space dedicated to innovation, a key component of which is the discovery of deep structures within the contents of our awareness. Latching onto this patterned, meaningful form of information processing is an immensely powerful way to learn, which accounts for why human consciousness has enabled us to take such great strides in every intellectual field we explore. By discovering the hidden rules in nature, by linking disparate ideas together according to their underlying common informational structure, we can weave a vast tapestry of meaning inside us. One consequence of this patient, piecemeal endeavor is that when we spot a chair, we don't see it according to its basic sensory features. Instead, we unavoidably recognize it *as a chair*, and immediately have access to a pyramid of meaning relating to this one object—what forms chairs take, what functions they serve, their relationship to other furniture, the rooms and buildings they inhabit, and so on. In fact, as we gaze around our world, our unconscious minds might be busily processing the basic sensory properties of each feature, but within the citadel of our awareness, we ineluctably view each component of the scene via the dense filter of the structure of knowledge we've acquired throughout our lives. Every single object on which we cast our eyes triggers a conscious wave of understanding, its own pyramid of meaning.

This roving appetite—not only for knowledge, but for profound patterns—is both the mechanism of innovation and the signature feature of human consciousness. DNA-based "ideas" cannot be conscious, partly because they are constrained to represent only the most basic facts of the world. Even chimpanzees struggle to understand hierarchies of meaning, but human consciousness thrives on this mental architecture, which enables us to understand and control the world with unique depth.

Consciousness concerns itself only with the most meaningful mental constructions and is ever hungry to build new patterns over existing architectures. To help in this aim, it itches to combine and compare any objects in our awareness. How the brain supports consciousness closely mirrors these functions. Those specialist regions of the cortex that manage the processing endpoints of our senses—for instance, areas involved in recognizing faces, rather than merely the colors and textures that constitute a face—furnish our awareness with its specific content. But there is also a network of our most advanced general-purpose regions that directly draws in all manner of content from these specialist regions. This is the core network, incredibly densely connected together, both internally and across major regions throughout the brain. In this inner core, multiple sources of meaningful, potentially highly structured information are combined by ultra-fast brain rhythms. And this, neurally speaking, is how and where consciousness arises.

We now have a sufficiently clear understanding of which brain regions are involved in awareness, how they communicate, and so on, to propose mathematical models of consciousness based on neural architecture and information signature, and indeed such models are already being proposed. Simultaneously, new empirical methods of indexing levels of awareness are emerging. The blistering progress underway promises answers to the previously impenetrable question of how we gauge the conscious levels of beings that aren't equipped to tell us about their consciousness, via language. This list of awkward subjects includes other animals, fetuses, babies, and—in the future—robots.

But there is another group of subjects who would call out even more pleadingly for a validation of their capacity for consciousness, if only they could. Patients with severe brain injury can appear to hover on the edge of awareness, with negligible behavioral signs that they are conscious. These patients' doctors may naturally conclude that the ravages of injury or illness have truly robbed them of the capacity for awareness. But what if the damage includes the brain's motor centers, leaving the patient utterly paralyzed? Are such patients secretly fully conscious, or has the extent of brain damage destroyed their capacity to experience the world, with paralysis therefore an irrelevance? How can we apply our scientific knowledge of consciousness to distinguish between these possibilities?

Our awareness gives us incredible gifts of understanding, though there is a heavy price to pay for such a vast consciousness. In the final sections of this book, I discuss the fragility of the organ that has grown so large and complex in order to support the amazing innovation machine of human consciousness. We are especially prone to serious brain injury, which can persistently rob us of awareness. Thankfully, though, many new techniques are arising to diagnose the levels of awareness that may still secretly reside in brain-damaged patients. Extensions of this research are beginning to offer us a chance to "hear" these patients, just by reading their brain signals, and for them to communicate with the outside world. Some emerging methods may even allow us to restore some degree of consciousness to patients in which it is clear that awareness is tragically absent as a result of injury.

Cases where severe brain injury leads to a persistent twilight of awareness are, thankfully, relatively rare. Unfortunately, though, the fragility of the human brain manifests very commonly, in more subtle forms. For optimum consciousness to occur, a complex interplay of various brain chemicals and activity between regions must be balanced just right. Some people have genes that make brain instabilities likely, and much of the population can be repeatedly battered by life's stressful events, which further strains their intricate neural machinery. The result can easily be mental illness, a pandemic that gets far less focus than it deserves.

But vital new clues in both understanding and treatment are arising, with almost all psychiatric conditions being repainted in terms of disorders of awareness. Some psychiatric conditions involve a deeply deflated conscious space, like a car that can only crawl on a dangerously icy road. Such patients are desperate to move through life at a more normal pace. They frantically cling to any meager hint of a pattern that they perceive, like a driver who assumes that the only solution is to slam his foot on the gas pedal. This panicked response causes the car to skid out of control. Likewise, the patient's mind freewheels, generating a multitude of paranoid, spurious innovations, which we call delusions.

Various techniques that literally expand and reinvigorate consciousness are being successfully applied to almost all psychiatric patient groups. However, this is not just the story of what consciousness is, and when it breaks down, but how we can apply this knowledge to aid our daily lives. For instance, many of these awareness building approaches could just as easily be adopted by all of us, both to reduce the daily weight of stress we endure and to enable us to view the world more directly, with fresh eyes. And, in time, we can learn tenderly to nurture a consciousness that is quiet, open, and ready to discover many beautiful new patterns around us.

Conceptual Conundrums
of Consciousness

Philosophy

TECHNOLOGICAL TELEPATHY

Many people share the easy intuition that our minds are somehow separate from our bodies. Most religions have encapsulated this notion via the supposition that we continue to live, in some form, after our bodies have perished, either in the afterlife or within another animal via reincarnation. Therefore, according to such theologies, the brain and body have nothing to do with our consciousness. But if this were really true, then how could an aspirin pacify my pain? How could that morning shot of espresso dispel my drowsiness? Is it really a remarkable coincidence that these drugs change my brain chemistry at the same time that specific aspects of my experience are altered?

If these examples of the physical world influencing our mental states are too subtle, then let me provide a more direct case. A few years ago I went into a brain scanner as a guinea pig for a friend and colleague, Martin Monti, who was trying out a new experiment. I'd been inside the scanner many times before—it is common practice for researchers to scan each other using rough-and-ready versions of a test before finalizing the experiment and bringing in volunteers from the general population. But this time was very different. This time, Martin was going to use the scanner as a telepathic tool.

Normally I would be performing a task in the scanner as I watched a series of images, and the complicated demands of imaging analysis meant that my brain activity could only be deciphered days later. But Martin was now using a newer procedure to carry out a far cruder analysis on the fly that would enable him to see my brain's activity mere seconds after it occurred. The resulting picture of which regions had lit up would be far less detailed than what we normally viewed, but for his purposes they were sufficient.

This particular study involved him asking me various questions. If I wanted to answer yes, then I'd imagine looking around my house, which would activate the navigation region at the bottom of my brain (the parahippocampal place area). If I instead chose to answer no to his question, then I was to imagine playing tennis, which would activate the part of my motor cortex responsible for my hand and arm—at the top of my brain. These two tasks were deliberately designed to produce well-documented, robust, but diametrically opposite activation patterns, thus making them amenable to crude real-time analysis. Such a pattern of brain areas lighting up works consistently between subjects because our brains compute such tasks in very similar ways.

The experiment was part of a project to attempt to communicate with patients who might be fully conscious, but are unable to show this to the outside world because they have lost all motor control. But before subjecting severely ill patients to the inconvenience of a scan, Martin needed to hone the technique on guinea pigs like me.

During most former occasions when I lay in the scanner, I had felt relaxed—even a little bored—and occasionally struggled to stay awake (not that I would admit this to the researcher!). Now, though, I was surprised to find how excited I was. As the radiographer pushed the bed I was lying on into the large, white, fattened-donut shape of the fMRI scanner, so that only my feet were outside, I realized I was even a little nervous. It almost felt as if I were mentally naked—that Martin could actually watch my thoughts from the console room as he stared at the scanner monitor. I knew that this wasn't really how it worked, but nevertheless I felt a palpable, exhilarating vulnerability. I was about to become more mentally transparent than almost anyone in the world had been before.

After the various loud beeps and clicks of the calibration scans, I heard Martin's voice in my earphones asking me whether I had any siblings. For the next 30 seconds I recalled the various pieces of furniture and the shape and size of rooms in my home: meaning *yes*. Then I could relax for 30 seconds (the control) before repeating this minute-long cycle another four times. After the 5-minute train of piercing scanner beeps had stopped, the brief silence was broken by Martin's voice in my headphones: "Okay—so you *do* have siblings. More than one?" Another 5 minutes passed as, in the non-rest periods, I imagined wild forehand swings with the tennis ball coming at me very fast: *no*. Then Martin's voice again: "Okay, so you only have one sibling. A brother?" More house-browsing for yes. Martin correctly deciphered every answer I gave, simply by staring at a computer monitor that was representing my brain activity. In fact, he could normally do this within the first minute, with the other 4 minutes of trials seemingly only there for reassurance.

Once the family questions had been exhausted, the conversation resorted to outright chattiness: "So do you think England is going to win the world cup match tonight?" Knowing little about soccer, except for the quality of the England team, I frantically started playing tennis to indicate: *not a chance*. We continued to have a conversation like this for about half an hour, with me thinking in these 30-second chunks, and Martin looking at the pattern of brain activity, and knowing very quickly whether I had answered yes or no. Admittedly, with a question every 5 minutes, it was not the most efficient conversation I'd ever had, but let me emphasize what was occurring: This conversation was being carried out without one of us engaging in any form of speech, gesticulation, or writing. I was answering in my mind by pretending to do various things, and Martin was detecting my answers by looking at my brain activity, *as I was thinking those answers*. When the radiographer helped me off the scanner bed and removed the various wires and equipment that had surrounded me, I paused for a moment and thought: *I have just participated in about the most definitive demonstration in existence that the mind is nothing more than the brain.*

It's not as if this were the first time I had believed this, of course, having been heavily swayed by the evidence of the personality changes my father endured when the right side of his brain was swollen and constricted. But even that didn't diminish the impact of what I'd experienced. I had successfully undergone science's equivalent of telepathy; Martin had watched the inside of my skull as if it were a film—right at the time when I was co-opting my imagination to project the right images onto this "movie screen" of my brain.

PHILOSOPHY VERSUS SCIENCE

This book is shamelessly about the science of consciousness. Every chapter except this one will explore the evolutionary background and psychological and neural mechanisms of our own experiences. But questions about the relationship between the mind and body have been fiercely debated in philosophical circles for well over two thousand years. In fact, only in the past two decades has there been a clearly visible consciousness research field. It would therefore be remiss of me in a book on consciousness to ignore the major philosophical debates, which are such a well-established ancestral influence on consciousness science.

I will firmly assert, however, that these philosophical arguments, which rely so heavily on abstract logic for ammunition, as they neglect the scientific enterprise, provide very limited insights into consciousness, and can be positively misleading.

I'll be centering on two key questions. First, is there nothing more to consciousness than brain activity, as my time in the scanner implied, or is awareness somehow independent of brains, bodies, and the rest of the physical world? And second, are we as mental beings nothing more than biological computers, or is there something special about the sensations we experience, and the meaning we attribute to the world, that could never be captured in software form?

DESCARTES AND THE MIND-BODY DUALITY

The seventeenth-century philosopher René Descartes is the landmark father figure of the philosophy of mind. In his most famous work, *Meditations on First Philosophy*, Descartes contemplated the possibility that a "malicious demon of the utmost power and cunning" had deceived him about the existence of all external things, including his own body. This is essentially also the premise of the film *The Matrix*, in which the hero, Neo, goes about his daily life believing that he is living in a twentieth-century U.S. city, only to be woken up from this extended dream to realize that it was a simulated reality that devious computers had been generating and wiring into his brain.

Descartes recognized, though, that no matter how malicious this demon was, there was one realm of thought that was certain, impervious to the demon's illusions: his own existence as a thinking being. Like Neo, you may believe in blissful ignorance that you have the same body you've always had, as that is what the computers feed into your senses. But the one act beyond the power of these evil computers is to fool you about your own existence. There are two options: If you *do* believe you exist, then logically you must exist—at least as some kind of conscious being—since the act of believing requires the existence of a conscious being to believe it. Alternatively, if you try, somehow, to believe you *don't* exist, then the very act of doubting confirms your existence again, since doubt also requires a conscious being to perform the doubting, as it were. Therefore, just by the act of thinking (with doubt as one example), you know that there must be a conscious entity around, and you also know that it is you!

In the meditations, Descartes articulated this idea as: "I must finally conclude that this proposition, *I am, I exist,* is necessarily true whenever it is put forward by me or conceived in my mind." But he put it more famously and succinctly in *Discourse on the Method* as "*Cogito ergo sum*" (I think therefore I am).

One of Descartes' main arguments for justifying the mind-body duality was intimately bound to these views on doubt. The argument was deceptively simple and superficially persuasive: Because we can so effectively doubt the existence of our own bodies, but can never doubt the existence of our own minds, the mind is completely distinct from and independent of the body (a modern spin on this argument might substitute "body" for "brain").

The brilliant philosopher, mathematician, and logician Gottfried Wilhelm von Leibniz, who was born around the time of Descartes' death, was quick to vilify Descartes. Leibniz pointed out that all Descartes had actually shown was that he *could contemplate* that his conscious mind was distinct from his body. He certainly hadn't proven anything. This critique can be illustrated by a slight twist on a well-known example. Say I happen to be walking the streets of Metropolis and from a distance I see a tall, well-built man with thick, ugly glasses hurrying into a telephone cubicle in an alley.

My friend tells me that he's the *Daily Planet* reporter, Clark Kent. Suddenly, on the other side of the street, five gunmen descend on a security van, looking to steal hundreds of thousands of dollars in cash. I'm terrified and excited at the same time, and believe they'll get away with it. Then I feel a momentary swirling wind, and miraculously, as if from nowhere, Superman flies past me toward the criminals. He has them disarmed and tied up in ropes in the blink of an eye. My friend looks at me, his head cocked sideways, and asks: "Do you . . . do you think there's any chance that this Clark Kent guy is the same person as Superman?" I laugh at how ludicrous that suggestion is, and quickly retort: "Listen, I definitely know who Superman is—I've seen him fly around loads of times. I've even interviewed him twice for my magazine. I barely know this Clark Kent guy, and besides, from my fuzzy glimpse of him a minute ago, he even looks different because he wears glasses. Therefore I'm certain that Superman and Clark Kent are two completely distinct people." My friend nods, impressed at my watertight logic, and I feel a warm, comforting sense of smugness at yet another example of my superior intellect.

The Superman observer is making two mistakes here: First, he's assuming that his own level of knowledge of Superman/Clark Kent *is an actual characteristic of Superman/Clark Kent*; second, he's assuming that superficial differences between Superman and Clark Kent must mean they are different people rather than two versions of the same person. *But* if he were a decent, professional reporter, seeking out definitive evidence like a bloodhound, his conclusions on Clark and Superman would be very different. If he studied Clark's bizarrely frequent visits to telephone booths and knew that Superman always popped out of the very same booths moments later, if he found out that Clark always wore a Superman costume under his normal clothes, if he saw what Clark looked like without glasses, and so on—it would be blindingly obvious they were actually one and the same person.

Descartes' argument, essentially based on his ignorance of the brain, is underpinned by a similar unwillingness to explore the evidence. To comprehensively test his claim, just as the bystander should be studying every detail of both Clark Kent and Superman, we would need to know everything about our brains and our awareness. If there are instances when consciousness radically alters, but brain activity is unchanged, then we can start talking about independence of brain and mind—*but not until*. As it is, all brain-scanning experiments to date have shown that even the subtlest of changes in consciousness are clearly marked by alterations in brain activity. The alternative perspective, then, that consciousness is a physical, brain-based process, is eminently more plausible than the belief that consciousness is independent of the physical world.

But Descartes also claimed that our minds are necessarily private, subjective, and unobservable by others. It's worth lingering on this point. When I look out at the vast ocean, hear the pulsing murmur of the waves, and feel a sense of peace and contentment, no one else will ever experience precisely what I experience at that moment. In an absolute sense, it seems that I really am trapped, alone, inside my head, and there's nothing science can do to change this. To extend Descartes' assertion in the modern world, brain scanners may capture an approximation of my consciousness, but could they ever, even in principle, enable someone else perfectly to experience what I just experienced? This question reflects the abiding mystery of subjectivity, which remains the inspiration for modern attempts to demonstrate the independence of mind and brain.

Finally, it is worth pointing out that Descartes, like everyone else who thought about such things until a century or so ago, assumed that the mental realm simply meant everything he was conscious of. Descartes would probably have viewed the concept of unconscious thoughts as an oxymoron, and certainly would never have accepted that our unconscious minds could influence our consciousness, as we all now largely assume. For the record, whenever I use the term "mind" from now on, or discuss "mental states," I'm including all cognitive processing, conscious or not.

Although Descartes had contemporary critics who essentially believed that the mind *was* the physical brain (most notably the English philosopher Thomas Hobbes), Descartes' mind-brain duality was largely accepted, even by philosophers, for centuries.

MODERNITY ARRIVES AND GHOSTS LEAVE

Despite Descartes' prominence, beginning in the mid-nineteenth century within the medical and fledgling neuroscience communities there was mounting evidence that a dualistic position was simply untenable. The most famous neurological case of this period was that of Phineas Gage. Gage was a foreman working in railroad construction in Vermont. One day, while he was helping to clear a volume of rock using explosives inserted into a hole, the gunpowder exploded prematurely. The tamping iron he was using shot out of the hole like a bullet. The frightening piece of metal was 3 centimeters wide, over a meter long, and weighed roughly 6 kilograms. It penetrated his left cheek, shattering the bone, then shot through his left frontal lobe, probably destroying much of the front part of the brain (see Figure 1). Finally it shot out through the top of his skull, eventually landing 25 meters away. Although obviously in shock and losing blood, amazingly, Gage remained conscious at the time and could speak within a few minutes. He was even able to walk un-aided, and he managed to sit upright on the short cart journey to the physician. He eventually made a remarkable recovery, with one prominent exception.

A landmark paper by his physician, John Harlow, described how, before the accident, Gage was well balanced, smart, sociable, responsible, and re-spected. Afterward, however, Gage became immature, regularly profane, dis-respectful, capricious, and seemingly unable to follow through with most of the dizzying numbers of plans he kept conceiving. In short, his previous friends believed he was "no longer Gage." Shockwaves rippled through soci-ety with the story that a man's personality could be so radically altered by brain damage. Although there was considerable controversy about the be-havioral details of this case, in the decades to follow, dozens of similar in-stances were reported where brain damage led to changes in personality or intellect. Science was slowly beginning to turn the conceptual tanker around toward the idea that the mind simply was the brain.

It wasn't until the middle of the twentieth century, though, that the most famous and biting attack on Descartes' dualistic position was mounted. It came from an English philosopher named Gilbert Ryle. In his seminal 1949 work, *The Concept of Mind*, Ryle described Descartes' position as a "philoso-pher's myth." Ryle pointed out that Descartes, in positing the independence of mind and body, was making a basic "category mistake." As an example of a category mistake, let's say a foreign friend visits me in Cambridge, and wants a tour of the university. I show her St. John's College, where I was based as a PhD student, with its beautiful covered Bridge of Sighs over the boats punting on the river Cam, and its majestic New Court, which rather resembles a wedding cake. I then take her through various other departments and colleges, but after a while she grows impatient and asks, "Okay okay, I've seen where members of the college live, where scientists carry out research and all that, but I thought you were going to show me Cambridge University!" What my friend fails to understand is that all these buildings and people *make up* the university. They are not independent of it in any way, but are subcategories of a larger category called Cambridge University.

For Ryle, Descartes was making exactly the same kind of category mistake with the mind and brain. Descartes was perhaps aware that various brain regions contributed to sensory processing, but he nevertheless believed that the brain had nothing to do with our mental life. Instead, as Ryle put it, Descartes believed in "the dogma of the Ghost in the Machine." For Descartes, this mind of mine is some mysterious ghost living inside the biological machine that is my brain. But there is no need for a ghost in the machine. The machine of the brain is all that's required for a conscious mind to exist.

Before I leave discussing Ryle, I would like to return to a subtlety of the analogy, because I think it highlights something very interesting. Although, of course, my foreign friend was wrong to assume that the colleges and research departments were irrelevant to the concept of Cambridge University, perhaps she wasn't *entirely* wrong. If I said "Cambridge University equals every building, student and staff now attached, or who has ever been attached to Cambridge University," some people might argue that I was being too reductionist or unromantic in turning an eight-hundred-year-old institution into a set of components. There is meaning living within the phrase "Cambridge University" that cannot be entirely captured by a mere list of its component parts. There is the image that the population has of the university, which, for instance, is exploited in literature and films, and which carries with it perhaps a traditional, formerly aristocratic aura. The students themselves interact in ways that make the university embody something very academically minded, perhaps even a little nerdy, which is an atmosphere that cannot easily be explained by examining the university's parts. In short, there are *emergent properties* to the concept "Cambridge University" that will never really be captured by a shopping list of items that Ryle would class as subcategories of the university.

In many domains, surprisingly sophisticated forms of knowledge can materialize out of the intricate combination of ideas at a lower level, and these can very much seem to be greater than the sum of those subterranean parts. One clear example of this is money. Studying the atomic properties of credit cards and coins will not generate very much understanding about how the world's financial system works. Instead, many economic rules can be seen as emerging from the social interaction of people wishing to buy and sell goods and services. They can scale up to fiendishly complicated levels, with few people being able to predict the 2008 credit crunch, and hardly anyone understanding exactly why it happened. Another fascinating example is that of ants. A single ant is a very stupid animal indeed, capable of only the most rudimentary learning. One might assume that if you have a colony of stupid ants, all you have is millions of stupid creatures. But something almost magical happens when the ants interact (largely by chemical signaling)—they develop incredibly complex behaviors. These include farming (humans *weren't* the first species to farm, by about 50 million years), complex nest-building, an intricate division of labor among seemingly identical animals, and about the only known non-mammal example of teaching, as one ant guides another to a food source, even pausing every so often to let its student catch up. This has led some to suggest that an ant colony should not be seen as a collection of ants, but more as a *superorganism*. Perhaps our global community of humans, now so tightly interconnected via the Internet, is another such superorganism.

Emergent properties are not the sole realm of animate objects. The laws of gravity are relatively simple, and yet the stunning spiral shape of our Milky Way arises out of them. The equations for fractals tend also to be just a handful of characters in length, despite generating shapes of seemingly infinite and quite unexpected complexity. (See Figure 2 for various illustrations of emergentism.)

When you have an object such as the human brain, which is the most complex lump of matter in the known universe, there is a good chance that various emergent properties will materialize there as well. I'm not for one moment proposing Descartes' immaterial ghost, and shudder at its unscientific and religious connotations. But within a scientific, physical-based framework, I endorse and will discuss the idea that the brain is much more than the sum of its parts, and that consciousness may be its most shining, fascinating product.

THE IMPENETRABILITY OF "WHAT IT IS LIKE"

By the middle of the twentieth century, philosophy had largely caught up with neuroscience in believing that the mind was equal to the brain, and that any thought or feeling was really a collection of brain cells firing away. In fact, with the advent of computers, and the acknowledgment that we were nothing more than biological machines, the equation of mind with brain soon mutated into the equation of mind with computer program. We happened, by the quirks of evolution, to have been lumbered with this particularly wrinkly, jelly-like computer to substantiate the program of our minds, but it didn't have to be this way; we could, in principle, have just the same thoughts with a "brain" made of silicon chips.

This theory of "mind as a computer, which accidentally equals brain," is the most widely discussed philosophical position about the mind held today. It is also the view that almost all neuroscientists assume by default. But that hasn't stopped some modern philosophers from attacking it from almost every angle.

The first doubt comes from the suggestion that the mind can be entirely reduced to the brain (or computer, or whatever other physical object one would care to mention). Descartes opposed the possibility of this reduction, assuming that there was something intrinsically subjective and nonphysical about the mental world. In 1974, Thomas Nagel, in one of the most famous philosophy papers of the past hundred years ("What Is It Like to Be a Bat?"), echoed Descartes' position in modern form. Nagel accepted that thoughts could be characterized according to their ability to cause other thoughts and behavior, and he was certainly not opposed outright to the idea that minds were simply brains. But he did think there could be a problem with this view. If you and I hear Shostakovich's *Tenth Symphony*, I can make a great stab at imagining what it was like for you to hear the music. Of course, I may be entirely wrong in my imagination, but I can at least generate a plausible guess as to what you experienced. I might even have a good go at imagining what my cat experiences when she hears the doorbell ring. We have similar ears mechanistically, and our brains' primary hearing centers also aren't entirely dissimilar. But if I try to imagine what a bat "see/ hears" when it uses echolocation to navigate, then I have no idea where to begin. Assuming that a bat is conscious, then our two consciousnesses seem totally incompatible. I can gain absolutely no knowledge about bat consciousness—at least not in the realm of echolocation.* And if I can't even imagine what it is like to sense with echolocation, what hope is there that I can get a foothold using any of science's tools?

This "what is it like" aspect of thought, Nagel claimed, was the essence of consciousness, and it posed a problem, in particular, for those wishing to reduce consciousness to a physical process in the brain. Nagel believed that if some animal was conscious, then it had to have a "what is it like?" aspect to it. Nagel did not state that it was impossible for us to understand what it was like to be a bat, although he did suggest that this barrier was a fundamental problem for science, one that it had to face with radically different, novel approaches.

Australian philosopher Frank Jackson took this position one step further and argued that it actually *is* impossible for science to explain mental states using only physical processes. His argument revolved around a thought experiment, which went something like the following.

Imagine that the year is 2412 and evil philosophers control the planet. One small coven of hellish thinkers calling themselves the Descartes Brigade hatches an idea for a cruel but potent experiment. The most celebrated scientist husband and wife in the world have just given birth to a beautiful daughter called Mary, but at the precise moment she's born, the Descartes Brigade kidnaps her and locks her up in a windowless black and white room. They bleach her skin white, and even cosmetically alter her irises to be black, along with her hair. They feed her black and white foods through a small black hatch in the white wall, and as she grows up they entertain and teach her on black and white laptops and monitors. The physical sciences have been completed by this stage, and it is possible to know everything about physics, chemistry, and biology, especially including the brain sciences. Mary has little else to do, and anyway, like her parents, she has an aptitude for and love of science, so she takes it upon herself to learn this completed science. By the age of thirty, after decades of diligent study, she knows absolutely

---

*Actually, some blind humans do learn a form of echolocation sufficiently advanced, they claim, to distinguish between the front and back of a parked car. For an interesting article describing this, see "Echo Vision: The Man Who Sees with Sound," by Daniel Kish, in the April 11, 2009, edition of the *New Scientist*.

everything about the physical world (stupendously implausible, of course, but let's for the moment assume it's possible), from the nature of all the subatomic particles to the activity of every brain cell that represents color vision in humans. The members of the Descartes Brigade at this stage know that their plan is coming to fruition and, finally, they unlock the door to Mary's prison, letting her wander outside for the first time in her life. Dazzled by what she sees, overwhelmed, overjoyed, she stumbles into a nearby garden and bends down to stare at a red rose. As she views the scarlet color she exclaims, in shock: "Before being released I knew every physical detail of how the brain generates consciousness, but now I know something new: *I know what it is like to see red.* This extra knowledge is something that the physical sciences could never capture. Therefore there is something *nonphysical* about consciousness!" At this stage, she collapses and suffers a terrible nervous breakdown, but the philosophers of the Descartes Brigade callously rejoice. They believe that their poor guinea pig, Mary, has helped them show that consciousness is at least partly nonphysical. The evil philosophy gang members end their pamphlet by boldly proclaiming that Descartes was right all along!

Although this argument has indeed been influential, it is not as watertight as it might at first appear. In some ways it suffers from the same problems as Descartes' argument. Descartes made the mistake of overreaching with his level of knowledge (because he didn't know the existence of his body with as much certainty as he knew the existence of his mind, he leaped to the conclusion that his body was distinct from his mind, even though he never actually established that this was the case). Here Jackson was similarly overreaching by making strong assumptions about what a complete physical understanding of the universe would entail. Lord Kelvin, one of the greatest physicists of the nineteenth century, is reputed to have proclaimed, as recently as 1900, that "There is nothing new to be discovered in physics now. All that remains is more and more precise measurement." The timing of this claim was somewhat comical: That same year, Max Planck initiated one physics revolution by introducing quantum mechanics to the world. Then, five years later, Albert Einstein followed with a set of his own revolutionary theories, including special relativity and the equivalence of energy and mass.

We have absolutely no idea what this "completed physics" will look like in four hundred years' time. In fact, startling revolutions could turn up at any moment to thoroughly embarrass anyone clinging to scientific dogmatism. A twenty-fifth-century portrait of the universe may well be far more bizarre than superstring theory or quantum mechanics, and it would be rather pointless to speculate on the details. It would also be foolhardy in the extreme to assume with certainty, as Jackson's argument above seems to do, that such future physical scientific wisdom could not include a complete explanation of consciousness.

Indeed, this is not the only argument that can be raised against Jackson's thought experiment. Suppose that Mary, bunkered down in her black and white room, actually has a wistful fascination for the flora and fauna that lie just outside her philosophers' jail. So when she's not studying the physical properties of neuronal circuitry, her hobby involves learning about the garden just outside. She hacks into a robot in the neighborhood, which happens to have some decent cameras for eyes. She steers it to her nearby garden and by this means finds out exactly what wavelengths of light are being emitted from the rose bush over the next few days. She therefore discovers what shade of red the rose is. Just to be certain, she practices more hacking skills on some nearby twenty-fifth-century flying brain scanners, which just happen to be a popular tool of the Big Brother government of the time. By this means, she sees the brain activity of all the passersby as they glance at the roses, and she infers easily that this activity corresponds to the experience of seeing red in every case. So before Mary ever leaves the room, she has incontrovertible, highly detailed knowledge from a range of sources that the roses in the garden near her room are indeed red. It's important to emphasize, therefore, that when Mary finally is released from her room, she doesn't necessarily suddenly discover that the rose is red—*she could already know this*. All she actually knows now that she did not know before is "what it is like to see" red. And this is a very strange kind of knowledge indeed. So what has she actually learned, if anything? Some philosophers suspect that she has not learned any new information whatsoever.

Because of this suspicion, a few critics of Jackson's thought experiment have argued that "knowing what it is like to see color" is really like an *ability* to gather knowledge rather than knowledge itself. (In fact, Frank Jackson himself should be included on this list, because he has since rejected his former argument, largely in favor of this idea.) Our color vision allows us to learn many useful things, such as when fruits are ripe, or, in more modern climes, to know when a traffic light indicates that we should stop the car. But this information relates to knowing *that* something is red. Knowing *what it is like* to see red is more abstract, and perhaps could best be described as *an ability to recognize red if we came across it in the future*; in other words, a rather specific ability to gather color information directly, without recourse to external machines such as cameras.

Nowadays, we have multiple ways in which to acquire color information using natural and artificial technologies, such as our eyes or a digital camera. But whether the source is our eyes or some fabulous feat of modern technology, all that really matters is that the information *is* acquired, rather than *the way* it is acquired. The information, the knowing *that* the rose is red, is independent of the tool by which that information is acquired. By contrast, "knowing what it is like" is dependent on the tool used to gather the information—in this case, Mary's eyes. Consequently, it isn't "knowing" at all—it is merely the ability to use the specific tool of our eyes in order to acquire knowledge. So it's at least plausible that Mary didn't, after all, know anything new when she was finally let out of her monochrome prison and saw the red rose. Therefore, consciousness can still be a purely physical event.

CAN A PROGRAM HAVE FEELINGS?

The other aspect of the standard model of consciousness is that it's not only a physical process, carried out by the brain, but also a *computational one*. Modern philosophers have taken issue with this stance as well, attempting to argue that there are unique characteristics to consciousness in its natural biological form, which means we could never be converted into some silicon equivalent.

One prominent attack on the computational view of consciousness revisits the "what is it like" aspect of awareness, which includes all of our emotions and senses. The argument claims that the existence of this vital aspect of experience proves that consciousness cannot be captured by a computer. You and I both know that strawberries are red and blueberries blue, but what if my inner experiences of reds and blues are your experiences of blues and reds? Arguments along these lines assume it's quite conceivable that we would neither behave nor think differently when faced with a fruit salad. Consequently, the software equivalent of our minds could pick any old values to represent the reds and blues—or even entirely omit this bothersome bit of code, and by extension the rest of our color vision and all other senses and emotions—without weakening the fidelity of the program. But if this defining facet of our consciousness is a mere irrelevance to its computational equivalent, then that is a step too far, and computers simply cannot represent consciousness.

However, when the scientific details are taken into account, there is something ridiculous in the idea that you can simply swap red with blue, and leave all thoughts and behavior otherwise unchanged. Our perception of something as "red" is generated not just from the wavelength our eyes pick up, but also the vividness of the color, its comparison with the surrounding colors, its brightness, the meanings and categories of colors, and so on—and all this interacts with our other senses and feelings in an incredibly complex network of information (just think of "the blues" as a form of music depicting a class of emotion). All this perfectly mirrors the architecture of the brain, which is an inordinately dense web of connectivity, such that changing one region may modify the function of many others.

Consequently, my red cannot be your blue because there is no single, independent class of experience as "red." The truth, instead, is that all examples of "what it is like" that you care to pick, from "burgundy" to "melancholy," represent rich information about ourselves and the outside world unique to this moment, crucially not in isolation, but as a network of links between many strands of knowledge, and in comparison with all the other myriad forms of experience we are capable of. In this way, far from being irrelevant, our senses and feelings, although undeniably complex, serve a vital computational role in helping us understand and interact with the world.*

----

*A marginally more believable version of this philosophical argument is to keep the comparison within the same person. For instance, imagine that I had some bizarre surgery to rewire my color brain centers so that what I used to experience as red I now experience as blue and vice versa. But my objections still apply here. Even without any surgery, my perception of red today will not be my perception of red tomorrow: My experiential history and all the other senses and feelings that occur as I see red will be different each time, as my brain will be. So there is absolutely no hope for this much more radical surgical change.

Can a Laptop Really Understand Chinese?

The most famous defense of the idea that there is something special and nonprogrammable about our biological form of consciousness is the Chinese Room argument, first proposed by John Searle in 1980. The main purpose of this thought experiment was to demonstrate the impenetrability not of feeling, but of meaning. Searle was keen to prove that a human brain could not be reduced to a set of computer instructions or rules.

To describe the thought experiment, we need to turn to another gang of philosophers from the year 2412, Turing's Nemesis. Restless and rebellious, these philosophers are prowling the streets of New York with an aggressive itch for a dialectic fight. Soon, they come across a group of Chinese tourists and decide to play a mischievous trick on them. They show the Chinese group a plain white room, which is entirely empty, except for a table, a chair, blank pieces of paper, and a pencil. They allow the Chinese to inspect every nook and cranny of the simple space. The only features to note, apart from a door and a naked lightbulb in the ceiling, are two letterboxes on either side of the windowless room, linking it with the outside world. One box is labeled IN and the other OUT.

The ringleader of Turing's Nemesis, a thoroughly devious person, melodramatically explains to the Chinese group that he reveres their culture above all others and believes everyone else in the world does, too. In fact, he's willing to bet a month's wages that these Chinese people can pick any random sucker on the street, place him in this room, and that person will show that he worships their culture as well, because he will be able to fluently speak their language via the interchange of words on paper through the letterboxes. The exchanges will take place with the Chinese people on the outside and the random subject inside the room. The Chinese are quick to take up this bet (even in 2412, although quite a few non-Chinese people speak Mandarin, only a small proportion can write in the language).

The Chinese tourists take their time and pick a young Caucasian man. He does not seem particularly bright. He looks a little bewildered as they stop him on the street and pull him over. The ringleader of the philosophy gang accepts the man and helps him into the room. Out of sight, though, just as the ringleader shuts the door, he hands the man a thick book. He whispers to him that if he follows the simple guidelines in the book, there's a week's worth of wages in it for just a few hours of work. This book is effectively a series of conversion tables, with clear instructions for how to turn any combination of Chinese characters into another set of Chinese characters.

The man in the room then spends the next few hours accepting pieces of paper with Chinese writing through the IN box. The paper has fresh Chinese sentences from a member of the Chinese group outside. Each time the man trapped in the room receives a piece of paper, he looks up the squiggles in the book, and then converts these squiggles into other squiggles, according to the rules of the book. He then puts what he's written into the OUT box— as instructed. He is so ignorant that he doesn't even know he's dealing in Chinese characters; nevertheless, every time he sends them his new piece of paper, the Chinese are amazed that the answer is articulate and grammatically perfect, as if he were a native Mandarin speaker. Though the young man does not know it, he is sending back entirely coherent, even erudite, answers to their questions. It appears to the Chinese that they are having a conversation with him. The Chinese observe in virtual shock that he seems, totally contrary to first impressions, rather charming and intelligent. Amazed and impressed, the Chinese reluctantly pay the bet and walk away, at least able to take home the consolation of a glow of pride at the universal popularity of their culture.

With the Chinese group out of the way, the Turing's Nemesis philosophers decide to keep their human guinea pig locked in the room a couple of hours longer. One of the Turing's Nemesis members does in fact speak and read Chinese, and he translates each of the paper questions originally asked of the man in the room into English. He sends each question into the room in turn. The written answers, this time in English, come quite a bit faster. Although they aren't nearly as well articulated as they were in Mandarin, they are somewhat similar to the Mandarin responses he had copied from the book. This time, however, the man actually understands everything that's asked of him, and understands every answer he gives.

Now, claims the Chinese Room argument, if the mind were merely a program, with all its "if this, then that" rules and whatnot, it could be represented by this special book. The book contains all the rules of how a human would understand and speak Mandarin, as if a real person were in the room. But absolutely nowhere in this special room is there consciousness or even meaning, at least where Mandarin is concerned. The main controller in the room, the young man, has absolutely no understanding of Chinese—he's just manipulating symbols according to rules. And the book itself cannot be said to be conscious—it's only a book after all, and without someone to carry out the rules and words in the book, how can the book have any meaning? Imagine if almost all life on the planet went extinct, but somehow this book survived. On its own, without anyone to read it, it's a meaningless physical artifact.

The point of all this is that, when the rules of the book are used to write Chinese, there is no consciousness or meaning in the room, but when English is written later on, and a human is involved, there *is* consciousness and meaning. The difference, according to Searle, is that the book is a collection of rules, but there is something greater in the human that gives us consciousness and meaning. Therefore meaning, and ultimately consciousness, are not simply programs or sets of rules—something more is required, something mysteriously unique to our organic brains, which mere silicon chips could never capture. And so no computer will ever have the capacity for consciousness and true meaning—only brains are capable of this, not as biological computers, with our minds as the software, but something altogether more alien. Searle summarized this argument by stating that "syntax is not semantics."

This argument—like all of the others I've described—may appear to be an unbreakable diamond of deductive reasoning, but it is in fact merely an appeal to our intuitions. Searle wants, even begs, us to be dismissive of the idea that some small book of rules could contain meaning and awareness. He enhances his plea by including the controller in the room, who is blindly manipulating the written characters even though he is otherwise entirely conscious. Perhaps most of us would indeed agree that intuitively there is no meaning or awareness of Mandarin anywhere in that room. But that's our gut feeling, not anything concrete or convincing.

When you start examining the details, however, you find the analogy has flaws. It turns out that there are two tricks to this Chinese Room thought experiment that Searle has used, like a good magician, to lead our attention away from his sleight of hand.

The first, more obviously misleading feature is the fact that a fully aware man is in the room. He understands the English instructions and is conscious of everything else around him, but he is painfully ignorant of the critical details of the thought experiment—namely, the meaning of the Chinese characters he is receiving and posting. So we center our attention on the man's ignorance, and automatically extend this specific void of awareness to the whole room. In actual fact, the man is an irrelevance to the question. He is performing what in a brain would not necessarily be the conscious roles anyway—that of the first stages of sensory input and the last aspects of motor output. If there is absolutely any understanding or meaning to be found in that room for the Mandarin characters, it is in the rules of that book and not in the man. The man could easily be replaced by some utterly dumb, definitely nonconscious robot that feeds the input to the book, or a computerized equivalent of it, and takes the output to the OUT slot. So let's leave the human in the room out of the argument and move on to the second trick.

And to understand the second trick, we must pose a fundamental question: Does that book understand Mandarin Chinese or not?

THE MOST COMPLEX OBJECT IN THE KNOWN UNIVERSE

The answer to this question may seem simple. Our intuition tells us that there cannot be any consciousness or meaning in this special room because the small book is a simple object. How could one slim paperback actually be aware? But the thought experiment's second slippery trick is to play with the idea that something as incredibly sophisticated and involved as language production could possibly be contained in a few hundred pages. It cannot, and as soon as you start trying to make the thought experiment remotely realistic, the complexity of the book (or any other rule-following device, such as a computer) increases exponentially, along with our belief that the device could, after all, understand the Chinese characters.

Let's say, for simplicity's sake, that we limit our book to a vocabulary of 10,000 Mandarin words, and sentences to no longer than 20 words. The book is a simple list of statements of the form: "If the input is sentence X, then the output is sentence Y." We could be mean here. Let's assume that the Chinese people outside the room are getting increasingly desperate not to lose their bet. One of them actually thinks he half-spotted the Turing's Nemesis ringleader slip some kind of book to the guy in the room. Another member of the Chinese group happens to be a technology history buff and has played on a few clever computer simulations of human text chatters from the early twenty-first-century Turing Test competitions.* He suggests a devious strategy—that they start coming up with any old combination of sequences varying in length from 1 to 20 words, totally ignoring grammar and meaning, to try to trick the person inside the room into silence. How big would the book have to be to cope with all the possibilities? The book would have to contain around 10^{80} different pairs of sentences.† If we assume it's an old-fashioned paper book, then it would have to be considerably wider than the diameter of our known universe—so fitting it into the room would be quite a tight squeeze! There is also the issue of the physical matter needed to make up this weighty tome. The number of pairs of sentences happens to equal the number of atoms in the universe, so the printer of the book would be running out of paper very early on even with the first copy! Obviously, it would be hopelessly unrealistic to make any kind of book that not only contained every possible sequence of up to 20 words, but also connected each sequence as a possible question to another as the designated answer. And even if the book were to be replaced by a computer that also performed this storage and mapping, the computer engineers would find there was simply not enough matter in the universe to build its hard disk.

Let's try to move toward a more realistic book, or more practically, a computer program, that would employ a swath of extremely useful shortcuts to convert input to coherent output, as we do whenever we chat with each other. In fact, just for kicks, let's make a truly realistic program, based exactly on the human brain. It might appear that this is overkill, given that we are only interested in the language system, but our ability to communicate linguistically is a skill dependent on a very broad range of cognitive skills.

---

*Currently, every year computer "chatterboxes" compete for the Loebner Prize, which provides a forum for programs to attempt to pass the Turing Test, and convince a sufficient proportion of ordinary people that they are text-chatting to a human and not software. Every example of software to date has been an obvious simulation, not anything anyone would consider conscious. Nevertheless, these chatterboxes can maintain surprisingly realistic conversations, thanks to some very clever programming. To read more and try chatting to a few of the winners yourself, go to www.loebner.net/Prizef/loebner-prize.html.

†That's a 1 with 80 zeroes after it, or, in words, a hundred million trillion trillion trillion trillion trillion trillion. If we hadn't put a limit on the sentence size, then the number of possible sentences would have been infinite.

Although almost all neuroscientists assume that the brain is a kind of computer, they recognize that it functions in a fundamentally different way from the PCs on our desks. The two main distinctions are whether an event has a single cause and effect (essentially a serial architecture), or many causes and effects (a parallel architecture), and whether an event will necessarily cause another (a deterministic framework), or just make it likely that the next event will happen (a probabilistic framework). A simple illustration of a serial deterministic architecture is a single line of dominoes, all very close together. When you flick the first domino, it is certain that it will push the next domino down, and so on, until all dominoes in the row have fallen. In contrast, for a parallel probabilistic architecture, imagine a huge jumble of vertically placed dominoes on the floor. One domino falling down may cause three others to fall, but some dominoes are spaced such that they will only touch another domino when they drop to the ground, which leaves the next domino tottering, possibly falling down and initiating more dominoes to drop, but not necessarily.

Although modern computers are slowly introducing rudimentary parallel features, traditionally, at least, a PC works almost entirely in a serial manner, with one calculation leading to the next, and so on. In addition, it's critical that a computer chip functions in a deterministic way—if this happens, then that *has to* happen. Human brains are strikingly different: Our neurons are wired to function in a massively parallel way. The vast majority of our neurons are also probabilistic: If this neuron sends an output to many other neurons, then that merely makes it more (or sometimes less) likely that these subsequent neurons will be activated, or "fire."

Why have one form of computer architecture over another? Partly because a serial deterministic architecture is so simple and straightforward. A computer can apply billions of very basic calculations in a second, whereas my brain carries out only a few major thoughts at a time, at the most. Consequently, it takes my PC a fraction of a second to calculate the square root of 17,998,163.09274564, whereas most of us would give up on such a fiendish task before we'd even begun. But because of the parallel, probabilistic nature of our brains, our processing is far more fluid and nuanced than any silicon computer currently in existence. We are exquisitely subject to biases, influences, and idiosyncrasies. For instance, if you read the following words: "artichoke artichoke artichoke artichoke artichoke," you will spend the rest of the day (or even somewhat longer) recognizing the word "artichoke" a little bit quicker than before (and you might even be a little more likely to buy one the next time you go to the supermarket). My word-processing program simply produces angry red lines to punish me for my ungrammatical repetitions of "artichoke." It does not learn to insert those red lines any quicker by the fifth repetition of the word, compared to the first.

This continuous, subtle updating of our inner mental world means that we can also learn virtually anything very effectively. For instance, we would consider the task of distinguishing between a dog or cat in a picture to be a simple and trivial matter. But computers are still cripplingly impaired at such processes. The reason is that although recognizing different animals appears basic to us, such skills are, in fact, behind the veil of consciousness, fiendishly complex, and ideally they require an immensely parallel computational architecture—such as the human brain. Of course, it makes no sense for evolution to have shaped our brains to be highly skilled at accurately calculating square roots. But, from a survival perspective, having a general-purpose information-processing device, which can learn to recognize any single critical danger or benefit in a moment, and then appropriately respond, is highly advantageous.

Therefore, over a few seconds, serial deterministic processing is best suited to performing huge quantities of simple tasks, whereas parallel probabilistic processing is only effective at carrying out a handful of tasks. But these tasks can be very complex indeed.

In order to capture the scale of the challenge ahead for anyone wanting to make a book or computer that could speak the Chinese language, we need to delve further into the details of our human probabilistic parallel computer and understand precisely how it differs from a standard PC. Assume for the moment that a single neuron is capable of one rudimentary calculation. There are roughly 85 billion neurons in a human brain. An average PC processor has around 100 million components: so, about 850 times fewer components than a human brain has in neurons—an impressive win for humans, but not staggeringly so, and indeed there are some supercomputers today that have more components than a human brain has neurons. But this is only the beginning of the story. There is another critical feature of human brains that, in a race, would leave any computer in the world stumbling along, choking pitiably on the dust of our own supercharged biological computational device. While each component on a central processing unit may be connected to only one or a handful of others, each neuron in the human brain is connected to, on average, 7,000 others. This means there are some 600 trillion connections in the brain, which is about 3,000 times more than the number of stars in our galaxy. In every young adult human brain these microcables, extended end to end, would run about 165,000 kilometers— enough to wrap around the earth four times over! The complexity of the human brain is utterly staggering.

To begin to understand the sheer vastness of the parallelism of human brain activity, imagine that the human population is around 85 billion people, about 12 times what it is now. You've suddenly discovered an earth-shattering revelation, and you simply have to tell everyone. You e-mail every single person in your contact list, all 100 people, and tell them to pass this wondrous insight on to 100 new friends. They do so, then the next group follows the same instructions, and so on. Let's assume for the moment that there are only a few overlaps in most people's address books, and, for the sheer, unadulterated genius of the wisdom imparted, that everyone obeys the instruction to forward the e-mail within a few seconds. From the starting point of a single send, it actually only takes six steps for the whole multiplied world of 85 billion people to get the message, and a handful of seconds. Indeed, in the human brain it's thought that no neuron is more than six steps from any other neuron in the family of 85 billion.

Now imagine *everyone* in the world having such a revelation and *everyone* e-mailing their 100 address-book contacts about the news each time—and *everyone* doing this about 10 times an hour. If you didn't turn on some fantastically effective spam filters, your inbox would receive around a thousand e-mails an hour. Everyone in the entire population each receives a thousand e-mails in that single hour, collectively amounting to 85 trillion messages.

But a neuron may fire 10 times a second, instead of per hour, and send its output to 7,000 other neurons, instead of 100. So a nauseatingly dizzying complexity occurs in your brain every single second, with hundreds of trillions of signals competing in a frenzied, seemingly anarchic competition for prominence.

This massively parallel web of neural activity simply is the propagation of information. In many ways, this spread of data by minuscule parts is unintuitive: A PC stores a single piece of information in only one location, and that location cannot store any other data; in stark contrast, populations of neurons—those, for instance, in the fusiform face area (FFA)—store as an ensemble many different faces, with each neuron only contributing a small fraction of each memory for a particular face, but humbly capable of playing its minute part in supporting hundreds or even thousands of face memories.

But for all these differences, the fact of the matter is that both brains and standard computers are essentially information-processing machines, and are secretly far closer cousins than at first appears. So whatever algorithm a brain uses to process information could be recreated, in principle, on a PC. Indeed, in neuroscience, there are already prominent computer models closely approximating the biological characteristics of a large population of neurons (in one recent case, a million neurons, with half a billion connections), and these are showing interesting emergent trends between groups of pseudo-neurons, such as clusters of organization and waves of activity.

THE CASE FOR ARTIFICIAL CONSCIOUSNESS

To return to the Chinese Room: If the mind is indeed a program, then it's clear that this "software" occurs first and foremost at a neuronal level. So, it is simply too great a task for a book to represent all the incredible intricacies of human brain activity—the interacting complexities are just too staggering.

No, if we are to have some artificial device that captures the computational workings of a human brain, it needs to be a computer. And it's not inconceivable that a computer in four hundred years' time could be fast enough to run a program to represent our massively parallel brains, with their hundreds of trillions of operations a second. Let's give this computer a pair of cameras and a robotic arm. The arm can manipulate the pieces of paper fed in from the IN box and write new ones for the OUT box. Now, if this computer was able to communicate effectively with a Chinese person, what does our gut tell us about what's happening inside? I think it would take a brave person to claim with confidence that this immensely complicated computer, with billions of chips and trillions of connections between them, using the same algorithms that govern our brains, has no consciousness and doesn't know the meaning of every character it reads or writes.

To reinforce this point, imagine that four centuries into the future, neuroscience is so sophisticated that scientists can perfectly model all the neurons in the human brain.* The most famous neuroscientist in China, Professor Nao, is dying, but just before his death he is willing to be a guinea pig in a grand experiment. A vast array of wonderfully skilled robotic micro-surgeons opens up Nao's skull and begins replacing each and every neuron, one by one, with artificial neurons that are the same size and shape as the natural kind. This includes all the connections, which are transformed from flesh to silicon. Each silicon neuron digitally simulates every facet of the complex neuronal machinery. It could just as easily do this via an Internet link with a corresponding computer in a processing farm nearby, such that there are 85 billion small computers in a warehouse a kilometer away, each managing a single neuron. But although it makes little real difference for this argument, let's assume instead that miniaturization is so advanced in this twenty-fifth-century world that each little silicon neuron embedded within Nao's brain is quite capable of carrying out all the necessary calculations itself, so that inside Nao's head, by the end, is a vast interacting collection of micro silicon computers.

Nao is conscious throughout his operation, as many patients are today when brain surgery is conducted (there are no pain receptors in the brain). Eventually, every single neuron is replaced by its artificial counterpart. Now, the processing occurring in Nao's brain is no longer biological; it is run by a huge bank of tiny PCs inside his head (or, if you prefer, all his thoughts could occur a kilometer away in this bank of 85 billion small yet powerful computers). Is there any stage at which this famous scientist stops grasping meaning, or stops becoming aware? Is it with the first neuronal replacement? Or midway through, when half his thoughts are wetware and half are software? Or when the last artificial PC neuron is in place?

Or, instead, does Nao feel that his consciousness is seamless, despite the fact that a few hours ago his thoughts were entirely biological, and now they are entirely artificial? It is entirely conceivable, I would propose, that as long as the PC versions of his neurons are properly doing their job and running programs that exactly copy what his neurons compute, then his awareness

*Whether or not this is practicable, efforts to reach this goal are already underway. For instance, in one project, the mouse brain is being mapped at a resolution of 5 nanometers, which is sufficient to capture the detail of every cell in the brain. See www.mcb.harvard.edu/lichtman/ATLUM/ATLUM_web.htm.

would never waver throughout the process, and he wouldn't be able to tell the difference in his consciousness between the start and end of surgery.

Now let's say that Nao goes into the Chinese Room with his new silicon brain intact. Any Chinese person that passes by would have a perfectly normal paper-based conversation with him via the IN and OUT letterboxes, even though his brain is no longer biological. And both the outside conversationalists and the newly cyborg neuroscientist inside the room would assume that he, Professor Nao, was fully conscious, and that he understood every word. Is this not formally identical to the rule book that John Searle had in mind when he originally presented his Chinese Room thought experiment? We could even, for the sake of completeness, swap the book for Nao. We tie Nao's hands behind his back and bring back the young man who was in the room before, but this time with no book. The young man would duly show Nao the Chinese characters that rain down on the floor from the IN letterbox, and then follow Nao's instructions for copying out reply characters and posting them in the OUT box. The young man would still have no clue what he was helping to communicate, of course, but *something* in the room would—namely Nao!

And, of course, if the Turing's Nemesis gang offered the passing group of Chinese people the bet that any local guy could speak Mandarin, the situation would quickly turn sour: The Chinese group would cry foul as soon as Nao was led into the room with their choice of Caucasian subject. Even if it were made perfectly clear that Nao had a silicon brain, the Chinese group would very likely no longer be interested in the bet. Nao would appear fully conscious when tested, both in his mannerisms and conversation—and that would be all the Chinese group would need to steer clear of this silly scam.

I'm willing to confess that my brain-silicon-transplant thought experiment rests on untested intuitions, just as the original thought experiment did. For instance, it may never be practically possible to capture every salient cellular detail of our brains so that they could be exactly simulated within a computer. But at least Nao helps to rebalance things, by suggesting that any appeal to the mysterious, special, noncomputational nature of our minds rests on naive assumptions about how the brain works.

In the end, the most famous attack on the idea that meaning can be found in computer programs, Searle's Chinese Room, doesn't seem to be all that convincing, mainly because it implied that the programming required for language communication was grossly simpler than it actually is. Instead, we must at least be open to the idea that our minds really are our brains, which in turn are acting as computers running a certain (parallel) kind of program. Consequently, we could, in principle, be converted into silicon computers at some point, where real meaning and awareness would persist. There is certainly no convincing argument against this view, and I happen to believe it is extremely likely that silicon computers could, in the future, be just as conscious as humans.

The fact of the matter is that we are deluged with the idea of conscious robots in plausible ways in books or on the big screen, and they are believable to us because they are shown to have stupendously complex artificial brains. When we watch Data from *Star Trek*, or many of the characters in *Blade Runner*, for instance, we have absolutely no trouble entertaining the possibility that artificially created beings could be conscious in very similar ways to us. Indeed, almost all of these characters live inside worlds where a common theme is the unjust lack of rights they receive as machines. As characters, these androids are at least as aware as the humans enforcing their prejudiced rules.

ERODING THE WALLS OF SUBJECTIVITY

Throughout this chapter, the one enduring philosophical idea, the one argument that seems most robust and least answerable by science, is Descartes' notion that our consciousness is inevitably subjective. We can never be certain what anyone else experiences, and vice versa. We can replicate any physical property in the world—for instance, manufacture an identical PC on every continent—but we can't replicate experiences, which are locked inside of the head of the single owner of those experiences. This assertion seems obvious and right, but does it also rely on a set of untested intuitions?

In Vernon, British Columbia, lives a large extended family. Within this busy household, there is a pair of four-year-old identical twin girls, Tatiana and Krista Hogan, who are in many ways like any girls their age. They can be cheeky and playful, and sweet and caring, and when they get tired, they are just as talented as any other four-year-olds at becoming fractious and demanding. What makes them unique is that they are joined at the head and brain, with one twin forced always to face away from the other. Crucially, they seem to have a neural bridge between each other's thalamus, one of the most central and important regions in the brain. Among other things, this area is a sensory relay station. Although no rigorous scientific studies have so far been carried out on Tatiana and Krista, the anecdotal reports are utterly tantalizing. For instance, if you cover one girl's eyes and show a teddy bear to the other girl, the unsighted child can identify the toy. If you touch one girl, the other can point to where her twin has been touched. Tatiana hates ketchup, while Krista likes it, so when Krista eats something with ketchup on it, Tatiana sometimes tries to scrape it off her own tongue. Occasionally one twin will silently sense the thirst of the other and reach for a cup of water to hand it to her conjoined sibling. Therefore, one sibling seems to be able sense the vision, touch, taste, and even desires of the other. Most remarkably, each of the siblings appears to distinguish between those experiences that belong to herself and those that belong to her sister—though on rare occasions, such as involving ketchup tasting, deciphering whether oneself or one's sister experienced an event can be a confusing matter.

This striking case could suggest that Descartes was not completely right about the perfect prison of subjectivity, for in this example, one person is indeed privy to the subjective experiences of another. How much further could you go, in principle, to merge your consciousness with someone else's?

Returning to Professor Nao, we could raise the intriguing question of whether subjectivity need be private or special at all. Imagine what would happen if a few other people had their own conversions into silicon form, so that each person had a collection of chips dedicated to capturing and continuing their brain activity exactly. Imagine also that the programming unique to each individual and the record of the activity of every silicon node for each person was stored for posterity on vast hard drives. This immediately allows for a recreation of the Tatiana and Krista situation, with the sensory input from one person being fed via computer linkup into another's mind. But perhaps you could go a lot further. Perhaps more than senses could be combined—if conscious thoughts were shared between computers, could you hear the thoughts of another? Could you mentally become a double person—or more than double? Once one's mentality is in digital form as a series of algorithms in the computer, and everything boils down to information, a host of possibilities arises for how that information could be shared, each one breaking—or, more accurately, expanding—the walls of subjectivity.

Perhaps it would even be possible for one person to have his silicon mind gradually, over a few seconds, turn into the mind of another, maybe a long-dead relative, to explore the personality of that other person, relive that person's experiences, become subject to another's belief system, and so on, all via a computer algorithm that morphed their brain simulation into that of another. This could last a minute or two and then they'd revert back to themselves, but with some vague memory of what they'd just experienced inserted back into their own silicon minds. It would be an incredibly unnerving experience to have everything about you—your personality and all your memories—dissolve and be replaced by someone else's for a short time. But this possibility indicates again just how effectively the solidity of personal experience could in principle be transformed into something more fluid. If you fully explore the idea that our minds could be merely a kind of physical computer, all kinds of possible scenarios open up.

Of course, I'm now also guilty of indulging in various wild thought experiments without fleshing out the details. But I'm simply trying to show that another seemingly watertight argument, that of the impenetrability of subjectivity, perhaps instead rests on weak intuitions, and that the alternative is plausible.

Human consciousness will appear inexorably subjective if we assume that consciousness is a mysterious entity, immune to the penetrative eye of science. But if we instead assume that consciousness is actually a process created by the biological computer of our brain, whose driving purpose is to process information, like any other computer around, then we can start to demystify both consciousness and subjectivity. If, following on from this, we are open to the possibility that we can make significant scientific and technological progress concerning consciousness, then who's to say that subjectivity will at some point no longer be an inevitable feature of consciousness, but an accidental component, and one that is potentially easily corrected in various ways?

At the end of the day, therefore, even this last remaining philosophical mystery may dissolve. Instead of being a permanent, impenetrable barrier to the scientific exploration of consciousness, subjectivity might only reflect our lack of deep understanding *as yet* of how our brains process information and our current lack of technological expertise in capturing and manipulating that information.

Ultimately, the philosophical arguments summarized in this chapter claiming to show that consciousness cannot exist in a physical computational brain fail, not only because they neglect the details of how the brain actually functions, but also because they rely on intuitions, even if they at first appear watertight. But while intuitions can be a useful starting point in many topics, they should never be the endpoint. I believe instead that provisional ideas should inspire scientific investigation, where more solid answers lie.

OUR INDOMITABLE SPIRIT

When I was a child, my father read me bizarre, fantastical bedtime stories with vibrant characters, invariably set on alien worlds. One obscure, ailing, tatty book that utterly transfixed me was *The Space Willies* by the British writer Eric Frank Russell. The subtitle of the book, *You Can't Keep an Earthman Down*, aside from capturing the plot of the novel perfectly, completely summarized, to my mind, what makes humanity so potentially incredible. For me, hidden in that one phrase was a surprisingly complex emotion: that of being unblinkingly positive, absolutely goal-focused, totally confident in one's ingenuity to escape the tightest of traps, and even relishing the chance to exercise that ingenuity.

The novel, admittedly, was somewhat contrived and no doubt was dated even in my childhood, but it's also so funny—and so well executed—that you hardly notice such failings. It concerns a chronically nonconformist army pilot, Leeming, who crashes a spaceship behind enemy lines. He is soon captured by his lizard-like enemies and interned in a prisoner-of-war camp, where he is the only human. The situation looks bleak for Leeming, but he has one trait that makes him far superior to his jailers: guile. Leeming soon hatches an ingenious, if improbable, plan. He begins to spread a rumor that he, like all Earthmen, has a secret, shadow-like, but ever so powerful and vengeful twin. He jerry-rigs a twisted piece of wire and a shabby wooden block and starts to hint surreptitiously that he can use this ultra-sophisticated device to communicate with his remote twin and call an attack at any time. Not only this, but he suggests that even the main allies of his captors have similar, secret doppelgangers, called "willies," who could turn nasty in the blink of a reptilian eye. At first his guards are skeptical, but then they start sending out spies, asking humans if their enemies "have the willies." Obviously the answers are a hearty assent, along with the optimistic conviction that their enemies will only have more willies as the battle continues. Following some beautiful finessing of the situation by Leeming, and fortunately timed catastrophes befalling the prison guards, these rumors slowly grow to such gargantuan proportions that his captors, for their own safety, do all they can to release him and send him back to Earth. Eventually, the whole enemy alliance collapses under the weight of this single rumor.

In a roundabout way, this story taught my younger self that in any apparently insoluble situation, human ingenuity can successfully forge a path through various imposing barriers. The history of the study of consciousness has represented this proud habit, but also highlighted other more frustrating aspects to our collective character. For centuries we have been overly influenced by viewpoints from many quarters defending the intractability of consciousness to science. As this chapter has shown, many modern philosophers have emulated this position, producing a multitude of arguments for why a scientific approach to awareness may be pointless. For much of the history of psychology, even scientists have jumped on this defeatist bandwagon and avoided anything close to the study of consciousness, assuming it was simply unavailable to experimentation. For instance, George Miller, one of the most prominent experimental psychologists of the past century, suggested of consciousness in 1962 that "we should ban the word for a decade or two."

Luckily, from about a generation ago, we have also had scientists who shared Leeming's personality. They cheerfully ignored the cries from their colleagues that consciousness was the most insoluble mystery in the universe, and plowed on with a positive, exploratory attitude—just for the hell of it, to see what they could find. Such stories have been repeated myriad times in the history of science, with unscientific conviction against our ability to understand a topic dissolving into fascinating scientific advance. But this time, the situation is unique; this time, the topic is the very heart of what it is to be human.

From now on, I'll be abandoning philosophy. Instead, I'll focus on the success story of the science of consciousness. I'll describe how that brave, curious leap into the unknown has produced a cornucopia of fascinating evidence for what consciousness actually is, and how the brain generates our experiences.

A Brief History of the Brain

Evolution and the Science of Thought

The First Lesson in Nature Is Failure

Soon after I started my PhD at Cambridge's Medical Research Council (MRC) Cognition and Brain Sciences Unit in 1998, the director of the department, William Marslen-Wilson, came into my office. A tall man with dark, slightly graying hair and a kindly face, he chatted amiably with me for a few minutes, welcoming me to the department, which I was touched by—aside from the fact that he kept calling me various wrong names (he turns name confusion into an art form). Then, as he turned to go, he paused at the door and, with a whimsical smile, said, "Remember, David, the first lesson in science is failure." I took little notice of this rather mysterious piece of advice until I carried out my first ill-fated experiment, when, sure enough, my first lesson in science *was* failure.

Failures are an inevitable part of the process of doing science. As scientists, we are professionally trying to track the truth. We need to explore many different options in a creative, directed way in order to inch closer to what's really occurring in nature. Quite a few of those ideas have to be wrong, particularly if you take the scientific community as a whole, with its millions of competing scientists, many with differing views.

Consider, for instance, that for much of scientific history, it was believed that the universe was bathed in an amorphous substance known as the *ether*. Even by the end of the nineteenth century there was near universal acceptance of the idea of a "luminiferous ether," a medium to support the transmission of light waves across the vast expanses of space. Around the turn of the twentieth century, meticulous experiments carried out by Albert Michelson and Edward Morley, along with theoretical work by Albert Einstein, made this notion of a ubiquitous supporting substance untenable, and we now see "luminiferous ether" as a quaint, extinct theory.

In fact, calling long-rejected scientific theories "extinct" is a more apt metaphor than it might superficially appear. The similarities between the scientific method and biological evolution are surprisingly close because of the common underlying theme of information. The scientific method is concerned with data almost by definition. But perhaps not so obvious is the fact that the progression of scientific thinking is an evolutionary process: a preexisting idea mutates unexpectedly into a profound new theory, which captures something deep about the world, and gathers popularity, but always in competition with an array of differing hypotheses. It will continue to survive only if the proponents of rival theories fail to explain the world more accurately or to convince the minds of the collective scientific community to bank on their ideas instead. In this way, various species of useful potential information about our universe may emerge, thrive, and eventually die out, as if they were real biological species.

A more surprising notion is that all life is itself an implicit scientific enterprise, albeit one that cares only for information relevant to survival and reproduction, rather than anything whatsoever that is intriguing about the wider universe.

Nature resembles science closely in its catalog of successes and failures to such an extent that, just as the first lesson in science is failure, so one could easily claim that the first lesson in nature is also failure. Of all the species that have ever existed, only about one in a thousand survive today. But much more than this, the whole mechanism for specifying the recipe for life is honed from natural experimentation to creatively store and refine a set of working hypotheses about the world.

The main thesis of this book is that consciousness simply is a certain kind of processing of information, especially information that is useful, that captures some pattern to the world. This chapter will provide the context to this argument: Consciousness didn't pop mysteriously out of the biological ether. Instead, it evolved, like almost everything else in nature, in an incremental way, and is intimately linked to the universal blind biological enterprise of accurately capturing useful ideas. Consequently, almost all features of evolutionary "learning" are mirrored in the computational details of the brain, and of the landscape of our conscious minds.

THE ESSENCE OF EVOLUTION

The standard theory of evolution is beautiful in its simplicity: Over billions of years, from a single common ancestor, there arose all the millions of wildly differing life-forms that ever existed and that populate the earth today. Such variety occurs partly because there are limited resources, and creatures need to compete to grab the choicest morsels. Some organisms consume others, so competition can deteriorate into vicious battles, both within and between species. The traits that keep an organism alive and help it to breed will flourish, whereas features that hamper survival and successful reproduction will, over the generations, slowly disappear from the population. It is this shifting of traits within a population and across the generations that eventually creates new species.

These traits are mainly determined by genes, individual instructions in the recipe for how to make an organism. A creature's genes are copied to its offspring, which will therefore closely resemble its parent. But these recipe instructions can sometimes by chance get misspelled, potentially creating a new variety of traits with each successive generation. These misspellings are not like typos in books, which are always wrong. Instead, misspellings, or mutations, in the genetic code will sometimes be beneficial. And if sexual reproduction is involved, rather than simple self-replication, the resemblance across generations will be far from perfect, injecting even greater fluidity into the transformation of life over time.

From this schematic of how all life evolved on this planet, there is a hidden agenda, namely, the blind "need" to accurately represent the relevant features of the world. This requirement is in some ways the essence of evolution, and probably helped create genes, DNA, and even life.

BREEDING CHEMICAL COMPLEXITY AND REBELLIOUS OFFSPRING

Although there is a paucity of evidence to support the specific details of the origins of life, broad general comments can still be usefully made. In whatever location gave birth to life, there would have been a rich chemical soup of molecules. A small subset of these might have been capable of making copies of themselves from simple chemical reactions. (A surprising range of self-assembling non-life molecules have in fact been discovered and even technologically exploited.)

Multiply this tapestry of chemical activity by hundreds of millions of years of random interactions, and there will inevitably be a large variability in the qualities of all these different types of self-replicating micro-objects. Some molecules will require less energy and resources to generate copies, and these copies will be more faithful versions of the original, and more robust to potentially dramatic environmental changes.

One or two chemical forms may by chance be stupendously, ruthlessly good at making faithful replicas. In a sense, this is the first purchase point for evolution, even though there are as yet no life-forms.

Now, rapidly, there will be a thinning out of possibilities—all inefficient non-life replicators will lose the race to the resources and disappear, and the superior chemical copiers will dominate. The new battle is between these thoroughbred survivors. The active fight for energy and chemicals, even at this early pre-life stage, is an evolutionary process, because the main ingredients are already present: a vibrant competition for limited resources, on a superficial level, between different forms of chemical objects—and, more essentially, between different "ideas" about how to maintain one's shape and make copies—which the chemical details of these proto-creatures encapsulate.

For instance, it's a "bad idea" to be a great replicator dependent on potassium abundance when there's usually none of it nearby. There's no point requiring sunlight to maintain your shape when your habitat is normally in pitch darkness. There's also, more generally, no point having a chemical makeup that requires huge quantities of energy to replicate when energy is sparse, especially if rivals are around. Success as measured by a burgeoning population in the primordial soup is predicated on maintaining a chemical composition that reflects or tracks the environment most closely—what resources *are* readily available, what's the best way to extract energy from the local world, what environmental changes are likely to occur that may threaten many chemical reactions and that one may need to be protected against, and so on.

(At this stage, I should stress, I'm not assuming any consciousness whatsoever in any organism except for humans—terms like "beliefs" and "ideas" are meant as a kind of shorthand to describe creatures that internally represent a certain informational perspective about the world, but without any requirements for awareness of those representations.)

In this pre-life arms race, these close analogies to micro-beliefs about the environment, as stored in the shape of a chemical self-replicating object, are critical for survival. So it's natural to assume that those objects that somehow represent the world more accurately, with greater detail, will carry an advantage. Indeed, the key reason that life might have evolved from simpler non-life equivalents is that non-life could not have developed the complexity of physical structure, or, very closely related to this, the extent of information storage, that organic life as we know it easily can.

Let me illustrate with a schematic example. Imagine there are three primordial copying objects, Alice, Beth, and Claire, all close to an active volcanic vent. Alice has stored the information that this precise location equals resources ad infinitum (perhaps by a strong chemical bond to the rock wall). When this particular volcanic vent becomes dormant, she degrades; she doesn't make a single copy. Beth's chemical components, instead, hold within them the "idea" that resources can potentially be found in multiple locations (perhaps by a chemical bond to the rock wall that weakens without sufficient heat, but strengthens again when another heated rock is chanced upon). When this particular vent becomes dormant, the lack of heat means she detaches herself from the rock and floats randomly until she's jostled against another hot rock, which allows for a chemical reaction to bond her to the rock surface. She is again close to heat and other useful resources, allowing her to make some copies of herself. But when this rock, too, becomes dormant, and there is no other vent nearby, she degrades. In a sense, Beth's structure is molding itself more closely than Alice's to the external data concerning where resources can be found—instead of the chemical equivalent of a belief that Alice holds that "this location is all that matters," Beth's concept is that any hot rock will do. Claire has a physical structure that reflects the information that heat equals resources, regardless of location (by chemically sensing and gravitating toward the nearest heat source—behavior probably too sophisticated for non-life). So Claire has a chemical form that most accurately shapes itself to the information about her requirements for heat energy, as well as how in the world to find this, and this gives her—and her similar offspring—a distinct advantage. She follows Beth to the second vent, but when this vent fails and Beth degrades, Claire directionally moves toward the next nearest heat source. Over time, in response to these dangerously intermittent vents, Claire-forms will be the only population that survives.

LIVING ON THE EDGE OF CHAOS

While Claire's more sophisticated, accurate "idea" would have caused her to be the dominant pseudo-life creature in her world, an even more successful way of responding to a changing environment is to update your ideas about it. Making true copies of yourself is important, but with such a dynamic world, where superior rivals or new dangers might emerge at any moment, being too fixed in your representations is dangerous. In this situation, exact copies of the originally superior chemical look doomed by their antiquated inflexibility. So some mechanism that can actually inject new creative ideas—in other words, that can "learn"—could potentially be very advantageous.

At this primordial stage, on the cusp of life, changing "beliefs" simply means making nonidentical copies. In other words, a family of proto-creatures needs to maintain a healthy balance between keeping useful knowledge and accepting that their world-picture could be better; they want their offspring to be faithful copies of themselves, but not *too* faithful. This loosening of the fidelity of the information is potentially expensive, because by chance many offspring will be inferior, perhaps just disintegrating at birth, or in other ways missing some vital chemical detail that enhances the chances of survival or replication. But it also raises the opportunity for some of the next generation to be an improvement on the model.

This tension between maintaining beliefs and injecting new ideas is a profound issue for any complex information-processing system, be it proto-life-forms, the neural interactions in our brains, or the scientific enterprise as a whole. Usually, though, a Goldilocks middle state, with chaos on one side and utter stability on the other, is the optimal way for any system to process information, and especially to learn useful new details about the world. This semi-chaotic activity is found whenever efficient information processing is required. It is probably the default state for networks of neurons, and it is one explanation for how complex thoughts in the human mind emerge from neuronal chatter.

A similar optimal balance between order and chaos exists in the scientific enterprise. There are cases, particularly in the softer sciences, where a prominent professor with a large ego—and a history of drawing in a large amount of grant money based on his well-established ideas—will do all he can to maintain these theories, including engaging in practices that are essentially unscientific and dogmatic. He may bend the rules to publish papers confirming his results, ignore experiments his lab carries out that contradict them, insist that his lab tow the party line, that those working under him always believe in his theory absolutely, and so on. He and his scientific progeny, his PhD students and postdoctoral assistants, may well be maintaining this viewpoint in the face of increasing evidence opposing it. For a while, due to his influence and personality, his theory may continue to flourish, but eventually it will be superseded, and his research staff will find it increasingly difficult to grab decent academic posts because of their long-standing defense of a scientific position shown to be wrong.

In a separate category are scientists who constantly generate outlandish ideas but are not particularly interested in testing them with carefully controlled experiments. Admittedly these rarely get past the PhD stage, but if they do, their careers always seem hampered by their overactive creativity.

The best scientists not only have the most respectable careers but also leave a lasting legacy of work, along with a new set of high flyers, who were former students. These renowned scientists are skilled at establishing successful theories and empirical results. But they are also quick to ditch these theories when the evidence racks up against them. They then generate new ways of perceiving the field—always with a *qualified* creativity.

The ability to settle on this healthy balance between stability and chaos is probably too much to ask of pre-life creatures, except for the most advanced—those on the cusp of life—because they would lack the complexity to support it. Specifically, for effective, flexible information processing skills related to survival and replication, you first need a means of storing many solid preexisting beliefs, which DNA, as I will discuss in the next section, is supreme at doing. You then need techniques for testing new hypotheses about the environment. In life, the main method for this involves creating a host of successful offspring subtly different from yourself, with a small proportion of those differences potentially being an improvement, reflecting useful novel innovations.

Let me illustrate the relationship between complexity and adaptability with another schematic example. Imagine you have 5 different words (analogous to different kinds of atoms within a proto-life object) by which to make up a sentence 5 words in length (analogous to a replicating chemical creature made up of 5 atoms). In each case, the sentence of 5 words gives you very little information. However, there are 3,125 possible different sentences you can make. This is a reasonable number, but in the face of an incredibly dynamic world, it is still potentially very limiting. Now imagine you still have 5 different words, but you can make up a sentence 100 words long (like a replicating chemical with 100 atoms in it). Each 100-word sentence potentially carries 20 times more information than was represented by the simpler creature with sentences of 5 words. A far more striking feature, though, is that, instead of 3,125 possible different sentences, there are now 8×10^{69}! Therefore, if the capacity to represent a greater variety of ideas is beneficial, the chemical object needs to be larger and more complex. Some chemical designs of equivalent size will be better than others at storing information and getting the balance of stability and flexibility correct. The specifics of the design, along with complexity itself, will provide further hooks for evolution to clasp onto.

Once a certain complexity was reached, the emergence of life itself might have been rapid, explosive, and almost inevitable. Candidate life-forms, emerging into a mode of effective learning, would have carried an overwhelming advantage over their simpler, less flexible rivals. These thoroughbred proto-life knowledge trackers would have been able to adapt, becoming ruthless at exploiting available resources and forcing all the more stable, less flexible alternatives to turn to dust.

Reaching such thresholds, and shifting into higher gear as a result of them, also happens in other contexts. For roughly 99.5 percent of the time that humans have existed, for instance, little scientific progress was made. But over the past four hundred years, with aids such as the printing press, education, and a critical mass of people seriously interested in science, actively discussing theories, and recording evidence, collective human learning—and scientific discovery—have dramatically increased.

Wetware

At some point, in small, simple steps, basic proto-life objects probably evolved into early life-forms made up of RNA, which is a close cousin of DNA. Compared to any natural non-life alternative we know of, RNA is an exceptionally efficient and flexible information carrier.

How does RNA achieve this? Like DNA, RNA is a long string of connected components (known as bases) of four different flavors, or letters. A "triplet" sequence of three letters is an important combination—it is the way that RNA letters spell words—in DNA/RNA language, all words are three letters long. Each word represents one of the twenty or so amino acids, which are cellular building materials whose combinations form proteins. And proteins are essential for almost all functions of every cell of any organism on the planet. A whole sentence of a sequence of amino-acid-denoting words is needed to instruct the cell to make a specific protein. A whole sentence is also exactly what a gene is.*

Compared to those primitive pre-life copiers, which could represent limited information within their simple molecular structures, RNA can instantiate many times more ideas. It does this by building multiple protein

---

*This system is an amazingly powerful and efficient way of storing information, bearing marked similarities to conventional computers, despite the apparently meager alphabet of four letters. This may sound terribly limiting for an information processing device, but you can in principle code for an infinite variety of things with it. A standard computer can manage with just two options: a 1 and a 0. These two alternative digits can nevertheless represent huge quantities of different types of information, as long as the sequence of these 1's and 0's is long enough. If there's just one digit, it can hold just two different possible pieces of information (2^1—either a 1 or a 0). If I have two digits, that's four possible pieces (2^2—00, 01, 10, and 11). If I have ten digits, that's increasing handsomely, to 1,024 possible states (2^{10}—0000000000, 0000000001, 0000000010, and so on up to 1111111111). RNA (and DNA) works in a similar way, just with two extra number types in addition to 1 and 0. These four RNA possibilities could easily be labeled 0, 1, 2, and 3, but instead are commonly labeled A, G, C, and U to reflect the names of the actual chemical bases that these letters stand for. These are adenine, guanine, cytosine, and uracil. These are identical bases to DNA, except that the U (uracil) is a substitute for T (thymine) in DNA.

Why does this code utilize three letter words? A triplet sequence of any three letters allows for sixty-four (4^3) different combinations, more than enough for each possible three-letter word to represent the twenty amino acids, while two-letter sequences would fall short at sixteen (4^2) possibilities. That's why DNA/RNA words to code for amino acids in the recipe for proteins are three letters long.

molecules—potentially thousands within a cell. And each protein could be a far more complex chemical construction than would ever be possible in a simple non-life copying object.

We are now dealing with a system capable of enormous complexity and flexibility, even if any change in implicit ideas can largely only arise from the random changes of the RNA code in future generations. Before, it might have appeared a stretch to discuss simple replicating non-life chemicals as representing ideas about the environment, because the information would be so minimal and so closely locked into the shape and chemical properties of the object (although this immature information-carrying capacity was still the critical feature that evolution acted upon to move from non-life to life). But now it should become clear that an RNA-based life-form, with its special code of letters, like the 0's and 1's on a desktop computer, and its software programs for making proteins, is carefully shaped by evolution largely as an information-storage device. There is also a vast potential for adaptation across the generations as evolution tweaks the sequence of letters in order to update the successful traits recorded in RNA—killing off those creatures with letters that do not capture the world well, and nurturing those with letters that reveal the best ideas. In this way, the genes are not only storing information, but, if viewed over many generations, also blindly learning about how best to live in the world.

But while RNA is a mammoth step toward life compared with simple replicating chemical objects, it has various drawbacks. As a molecule, RNA is unstable and tends to degrade relatively easily. This is no problem for a short piece of information, which can be replicated quickly, but for anything longer, with many thousands of letters of information, it simply isn't practical. The longer sequence of information would deteriorate so quickly that the organism would have little chance of passing on to the next generation those useful qualities that natural selection bred into it.

In other words, if you want to increase your information capacity, RNA is not your molecule of choice. It simply doesn't scale up well: The more information it stores, the less information it can successfully pass on to the next generation. Any useful balance between stability (maintaining a belief) and chaos (creatively exploring new ideas—some good, some bad) will slide disastrously toward the chaotic side, and all beneficial concepts accumulated in that family of RNA life-forms will eventually be lost, inevitably along with the life-forms themselves.

DNA solves this problem. Bacteria, arguably the first real life-form, may seem to us exceedingly simple. However, even the smallest, most basic bacterium requires a DNA string of more than 100,000 letters of code in order to form the recipe of its biological makeup. DNA, despite requiring considerably more energy to copy, is vastly more stable than RNA, which means that far fewer mistakes appear when it is duplicated. For these reasons, there may well have been a strong evolutionary pressure for life to start using DNA as the primary storage molecule for information (with RNA now playing an intermediary role between DNA and protein). So DNA may have arisen relatively easily and early, especially since it is extremely similar in structure to RNA—the main difference being that DNA is made up of pairs of letters in a double strand, rather than the single strand of RNA.

I can now return to the issue of the extent of complexity and adaptability in a concrete way, asking these questions for life rather than for some simpler non-life alternative. When compared to the 3 billion letters of code in the human genome (the entire complement of genes in an organism), the 100,000 letters in a bacterium is tiny. Nevertheless, it is sufficient in principle to generate vastly more possibilities of different types of proteins than there are atoms in the universe. In fact, to exceed the number of atoms in the universe, 10^{80}, you only need a few hundred DNA letters. So bacteria can, in principle, be reprogrammed to do almost anything you can conceive of. For instance, biotechnology engineers are currently creating novel forms of bacteria to make diesel fuel as a waste product.

A Universal Recipe and a Universal Language

We have now arrived at the common blueprint for all life on the planet: DNA stores the instructions for the organism's structure and function, and RNA, having made a temporary copy of sections of this DNA code, in turn converts this information into many different types of proteins—and proteins are the key molecular tools of all organisms. This model must have been a wild success when it first occurred, by its dominance wiping out all the alternatives. Scientists believe this is the case because virtually all life that we know of, from the simplest bacteria to all the animals, including ourselves, shares exactly this process, and exactly the same language.

This universality of DNA is completely staggering. I recently stayed in a hotel on the east coast of Italy, in a village called Vasto, where I was forced to communicate with the local staff in a combination of incredibly meager Italian and awkward hand-gestures. It wasn't so bad, as I know a smattering of Spanish and French, and we all fumbled our way through conversations. All these languages are relatively similar to each other anyway. After all, we live on the same continent and are—of course—the same species. And yet, it's entirely understandable that the differences in our histories and culture, stretching back a couple of millennia, would have created changes sufficiently deep that most of the words in Italian sound and look at least somewhat different from English.

But the contrast between European languages and the language of genes couldn't be more extreme. Our evolutionary path diverged from bacteria a billion years ago. *Nevertheless*, the meanings of virtually every single one of the sixty-four possible words (these triplets of bases that code for amino acids) are identical between us. So throughout the biological world, it's not just that there is only one universal language, but that there is a single dialect of a single accent within this single language!

There are many practical examples of the extent of this staggering uniformity. For instance, scientists have countless times successfully spliced genes from one species into a completely different one and changed its characteristics. This includes human genes introduced into mice, and mouse genes into flies. You might think that such gene swaps are unnatural, the Frankenstein-like artifacts of biology labs. This couldn't be further from the truth. For instance, in the wild it is now assumed that about 13 percent of all plant species were formed by the melding together of two or more distinct lineages. And there are also well-documented examples of useful gene swaps between humans and viruses or bacteria.

Part of the reason for the rapid rise of this DNA-RNA-protein system, and its consequent total dominance, may well be the fact that it is very close to an ideal biological system for storing and processing information. DNA is a tremendously safe, reliable holder for vast numbers of genetic ideas, while the machinery in the cell can efficiently retrieve those ideas and express them as proteins. There is also a simple way of making copies of the entire organism: by unzipping the twin strands of DNA code, each strand making a single copy of itself, and then rejoining these into two double strands for the new cells.

With this ideal ability to store information and easily convert it into useful protein tools, the standard DNA-based mechanism we know today must have trounced the alternatives 4 billion years ago, and once this recipe for life took hold in the world, there was no going back.

Extra Innovation in Desperate Times

Now, though, with DNA-based life carefully designed to preserve large sets of ideas, it seems the scales have been aggressively tipped toward the boring, stable side of information processing, with little chance to adapt the DNA code when circumstances change.

To compound this problem, there are various biological mechanisms that make every effort to ensure that organisms avoid the chaotic route. For instance, the DNA words that code for individual amino acids can sometimes still be read correctly, even if they are slightly misspelled, because there are in many cases a few spelling variations for each word. There are also more active and sophisticated mechanisms in play in some organisms by which a careful proofreading of the code is done to detect and correct any errors as they arise.

But to even the score against DNA inflexibility, there is a second set of biological tricks. For example, whole sections of DNA can be mixed up to inject measured levels of chaos into the ordered DNA code, allowing a family of organisms over the generations to soak up new ideas.

Some fixed, equal balance between stability and chaos is an effective learning system: Now you can both maintain your DNA ideas and modify the code across the generations, so that new concepts can be gleaned from the world. But never deviating from this informational midpoint introduces its own inefficient, unintelligent stubbornness. In one extreme, possibly applying to a handful of bacterial species on the planet, if you are living relatively easily in a place that never changes over the millennia, with no enemies to speak of, then doing everything you can to keep your current successful ideas *just as they are* across the generations makes perfect sense. At the other end of the spectrum, if a species' genetic set of ideas are obviously unsuccessful, and many creatures are dying in droves, possibly because of a violently changing environment or many forms of competitors, then a mode of maximal innovation is the only likely road to safety. This is true despite the fact that the chaotic route itself will also introduce poor ideas that lead to even more deaths, since as long as new accurate genetic ideas are found, some members of the species will definitely be saved. In both cases, the middle ground is far from ideal.

In the usual niches that life inhabits, there is a mixture of good times and bad. Here, the ideal computational solution is to be able to tweak the ratio between existing beliefs and new ideas according to the circumstances. So, ideally, an organism should suppress any chaotic changes to its DNA in successful times, but then positively encourage such dangerous innovations when previous beliefs no longer work in a new, life-threatening world.

In a close analogy to this, the distinction between heavily grooved beliefs and innovation is one of the most prominent psychological features of human experience. We all have habits we've carried out a thousand times before, such as having that morning shower. I for one tend to spend the vast majority of each shower daydreaming, as my muscles unconsciously take over the tedious task of sponging myself clean. But if the water would suddenly turn stone cold, I'd come back to myself, know that something is wrong, and explore how to fix the faulty plumbing. This acknowledgment that an error has crept in, and that innovation is required to fix it, makes me feel more conscious, more energized, and certainly cuts out any daydreaming. In fact, it's no coincidence that moments like these initiate a spike in my awareness. I will elaborate throughout this book that this drive to innovate your way out of a problem is a crucial feature of consciousness, whereas, in contrast, an important role of unconscious, habitual plans is to implement those fully learned products of the initially conscious innovations.

Although simpler DNA-based life, such as bacteria, do not have any form of consciousness by which to modify their levels of innovation from dogmatic autopilot to desperate creativity, they nevertheless have an impressive suite of mechanisms by which to slide the level of creativity back and forth as they track the level of dangers in the environment. This provides striking evidence that sophisticated learning strategies, mirroring those that distinguish between conscious and unconscious thoughts, occur even at the humble level of single-celled organisms.

MUTANTS, SEX, AND DEATH

The first, most obvious means of injecting new ideas into DNA code is through mutations. Although DNA is an immensely robust molecule, DNA machinery isn't perfect, and occasionally errors do arise. For instance, for bacteria, one mistake occurs every 10 million letters. If you only have 100,000 letters to spell your entire recipe, that means there will only be a single mistake in a single letter for every 100 bacteria. Some of these misspellings won't even make any difference, as they will just be a new spelling of the same word. Others could radically alter the protein made—probably causing serious problems for the cell's functions. But there's also a slim chance that it will be an improvement, a better idea for how to survive and reproduce in the current environment.

With mutations being the mainstay of innovation in all organisms, manipulating this mutation rate is one way that creatures can increase the frequency of potential new ideas in order to match a more volatile world. Some species do indeed utilize this trick: When the situation looks grim, and survival is strained, random mutation rates are increased in some bacteria. Yeast react to stress not by reshuffling letters, but entire chromosomes, for the same inventive result.

An interesting analogy to this is in primate innovation. Those primates with the lowest social standing tend to exhibit innovative behaviors far more often than their higher-ranking compatriots, in the hopes of chancing upon some strategy that will raise them up the social ladder. There are many human analogues to this, such as the technological leaps that tend to occur in or around wartime.

Animals, however, with similar mutation rates to bacteria, but a far greater investment in complexity and size, have a serious problem: Since they reproduce up to half a million times slower than bacteria, their genetic creativity has taken a massive hit. This makes many animals terribly vulnerable to certain changes. The 10-kilometer-wide asteroid that crashed into the earth 65 million years ago was devastating for many animals, especially the dinosaurs, partly because they couldn't adapt fast enough to the climate changes it brought. Seventy-five percent of all animal species were made extinct by this event. Although it's impossible to collect such ancient data for bacteria, their extinction rate would very likely have been a very tiny fraction of this. Evolution would have been spoiled for choice to pick new forms of bacteria within most species, as they would have quickly adapted to thrive in the hellish conditions that arose after the asteroid's catastrophic arrival.

To attempt to compensate for this serious limitation (slow replication), animals reproduce sexually. Sex is in many ways the first port of call for new strategies. Although bacteria normally simply divide, preserving every gene in the process, they can also perform an analogy to sexual reproduction by combining with another bacterium, even of another species, and swapping a section of genetic code with their ephemeral lover. But for animals, sexual reproduction has to be very much the rule, rather than the exception.

From a "selfish gene" point of view, indulging in sexual reproduction, instead of simply cloning oneself, is a minor disaster, since only half of an animal's genetic identity is passed to the next generation. But the reward—genetic creativity—is very much worth it. Heavily mixing an animal's genes with its partner's throws up new genetic ideas in their offspring, helping them cope with the world's many threats. This compensation for slow reproduction is so useful that almost all animals exploit it.

One animal has been definitive in demonstrating the utility of sexual reproduction, the lowly nematode worm. One nematode species, *Caenorhabditis elegans*, is a favorite model of genetic research. Because these worms are very simple animals that rapidly create offspring (every four days or so), the case for sexual reproduction is marginal. *C. elegans'* response to this is to keep their options open, so they can either reproduce on their own or have sex with others.

From an information-processing point of view, if the worm's world is a safe paradise, replete with abundant, choice morsels, it may as well reproduce asexually, since its genetic ideas about how to survive in the world are accurate and successful. But if there are mortal dangers, then its DNA could do with a shake-up for the next generation, of the kind that sexual reproduction can offer, to see if its rather different children will chance upon a better genetic recipe to cope with this harsh world. In fact, this is exactly how *C. elegans* behaves. Patrick Phillips and colleagues have shown that, when faced with some threat, such as a bacterial infestation, these worms are more likely to forgo the default of self-fertilization and instead have sex with others, and because of this, the family line is more likely to survive. The cauldron of sexually induced genetic diversity is beneficial at those times. In contrast, any that are forced to self-fertilize, despite the same threats, simply cannot cope, and they are soon wiped out after only a handful of generations.

Another injection of creativity into evolutionary hypothesis testing may well be death itself. Some people believe that research departments benefit from forcing crusty old professors to retire at sensible ages, so that stubborn, old-fashioned theories and habits aren't perpetuated so forcefully in the community, and new ideas from younger, more dynamic scientists have more space to flourish. Likewise, in nature it's possible that the existence of death helps species to avoid the buildup of outdated hypotheses. It's true that organisms just wear out. It's also true that any fatal genetic illness that materializes after the creature has successfully had children is not something that evolution is particularly interested in removing. But this isn't necessarily the whole story.

For instance, death can be held at bay, seemingly indefinitely, in some cases. Some bacteria can survive, in stasis, in the cold wasteland of the Antarctic, for hundreds of thousands of years, if necessary. What's more, all organisms so far tested, from yeast to worms to humans, can, on average, have their lives extended by at least a third simply by eating less. It's therefore quite possible that this is an important biological mechanism by which to hang around for longer, until food becomes plentiful again and the environment is ripe for babies once more. So death, to some extent, seems programmed and flexible, and possibly for good reason.

I would speculate that without age-related death, genetic creativity across a species would become increasingly polluted by outdated ideas. If an older generation persists, then its offspring with genuinely useful innovations are less likely to flourish, as they have greater competition from their own family.

If this situation continues for many generations, then the good ideas will increasingly become diluted and the species will be far more sluggish in response to changes. And when some crisis looms, for which the creatures with this excessive longevity have no solution, the species will be far more fragile than it would have been with a rapid turnover of creatures across the generations.

A similar reason exists for why we don't, as a rule, remember everything we experience in our lives. Holding on to an increasingly irrelevant bank of information would drastically interfere with our daily functioning, and we would eventually be mentally crippled. One particularly striking case of near perfect memory is that of Solomon Sherashevski. Sherashevski was born around 1886 and grew up in a small Russian Jewish community, eventually, in his late twenties, ending up as a journalist.

It was as he began this profession that Sherashevski's extraordinary mental skills were revealed to the outside world. His editor was having his usual morning meeting with the staff to portion out all the instructions necessary for the reporters to go about town to do their daily jobs. Everyone was industriously taking notes—except Sherashevski. He, in stark contrast, didn't even have a pencil and paper at the ready. Assuming that Sherashevski was being lazy, the disgruntled editor called him up on his behavior. Sherashevski explained that he didn't need to take notes, as he simply remembered absolutely everything, all the time. Disbelieving, the editor asked him forthwith to prove this wild assertion, which he duly did, by quoting back with perfect fidelity every word that the editor had said that morning.

In fact, to this remarkable man, it was incomprehensible that other people *didn't* do exactly the same thing—*why on earth would someone immediately forget these important facts? What's the point of that?* At this stage, it was clear to outsiders that Sherashevski was far from normal. He was soon sent to a famous Russian psychologist, Alexander Luria, who studied him extensively over a period of thirty years.

Sherashevski's memory was indeed incredible. He seemed to remember almost everything he came across entirely naturally. One example involved him being read aloud some stanzas of Dante's *Divine Comedy* in its original Italian—a language he had no knowledge of. When given a surprise test on this content *fifteen years later*, he could recall the stanzas so completely that he even repeated the words with the same stresses and pronunciation as they were originally spoken to him.

Although the ability for such vast, faithful recall seems a fantastic mental gift, there were prices to pay, both big and small. One drawback of his exceptional recall was an occasional inability to see the forest for the trees, to discover meaning, structure, or patterns in the stream of information he was busy encoding. For instance, while he could memorize long sequences of numbers, he would be completely oblivious to any simple structure within them, such as ascending numbers 1, 2, 3, 4.

But these unfortunate quirks of his mind were nothing compared to the emotional consequences of his superlative memory. For instance, his imagination was so vivid, so complete, that he often would mistake reality for a daydream. At the very least, imagination would corrupt reality so profoundly that he would struggle to get through something as mundane as a novel—every word in it would conjure up too many distracting images. For similar reasons, he struggled to overcome the crushing weight of his past. Sherashevski claimed to have near perfect memories from before he was one. These were so striking that he fought in vain to banish these carbon-copy recollections, since they also included the overwhelming intensity of these earliest feelings—the absolute terrors, or racking sobs of infancy.

As Sherashevski aged, the burden of this enormous memory became increasingly difficult to endure. He became desperate to find some effective strategy by which to forget things. He drifted from job to job, and unfortunately he died believing that he'd somehow wasted the mental opportunity he'd been given, and that he had never really amounted to much.

Examples like Sherashevski demonstrate that sometimes the fading and death of old information can help a person succeed. The person who forgets an optimal amount of old material can have a more accurate, organized view of what's relevant in the world right now. Likewise, perhaps the death of older creatures can lead to a family or species with collective tools that are better honed for an ever-changing environment.

Replication has always been the driving force of evolution, with survival taking a back seat. But more than this, death as an evolutionary strategy might even be an example of how survival and replication can come to loggerheads, with replication not hesitating to abandon survival if there are gains to be made in terms of having a more accurate, up-to-date implicit picture of the relevant features of the world.

EVERY CREATIVE TRICK IN THE BIOLOGICAL BOOK

Mutations, death, and sex are by no means the only methods for potentially invigorating DNA sequences with useful new ideas, or tweaking the learning rate to reflect whether the microbe's current world picture is successful or deeply flawed.

There is a large array of tricks that various simple organisms can exploit to discover new ways to successfully survive and reproduce, but one of the most intuitive is simply to try moving a sequence of code somewhere else in your recipe. After all, if much of this code is capturing something useful—perhaps it already creates a functional protein—then its shift to another part of the genome could create a similar protein that might be even more beneficial. So, compared to making changes in a painstaking way, letter by letter, this method is both more powerful and efficient: The potentially useful idea is already half-baked. Of course, mixing up code like this could be utterly disastrous, but there is also a chance that it might be not just a step, but a great leap in the right direction of advantageous innovation.

This mixing up of whole nuggets of ideas happens in various ways in the DNA code. Entire sections of genetic code (called transposons, or "jumping genes") can jump around the genome, breaking off from one location and reattaching to another.

The source and behavior of these bouncing clumps of genetic letters is fascinating. Some of these jumping pieces of DNA might simply be a kind of life within a life—a ragbag collection of genetic letters that has chanced upon a way of surviving and reproducing, sometimes entirely within the dense, tangled forest of DNA strands. Just as an organism evolves through the generations, stumbling upon better beliefs about the environment, so are these jumping pieces of DNA code shaped by evolution—though their world is the cramped home of the set of DNA letters. But this isn't the entire story: If these leaping sections of genetic code cause catastrophic failure in the function of their host, the organism, then they, too, will cease to exist. At the same time, if they can in any way aid their host, then they will have a greater chance of survival themselves.

Most of these micro replicating machines (as Richard Dawkins might call them) are in fact remnants of viral invasions. If they were officially classed as a life-form, viruses might well steal the crown as the most supremely successful types of organisms on the planet. There are more viruses than all organisms put together. Viruses have probably existed as long as life has. They can only replicate when infecting a host cell, and because of this they are not normally classed as organisms in their own right. We have an ambivalent relationship with bacteria: We know that, although some may harm or kill us, we need them in our guts to digest food. Indeed, we have what were once bacterial invaders in every living cell, in the form of mitochondria, to supply us with energy. We resolutely loathe viruses, though—all they seem to do is make us ill. But perhaps we shouldn't be so hasty. A picture is emerging to show that viruses, too, may have their benefits.

Viruses, even smaller on the whole than bacteria, are also the most diverse and therefore the most creative forms of replicating machines on the planet. Their adaptability allows the flu virus, for instance, to infect us year after year, while antibiotic resistance in bacteria, the more boring, less flighty cousin of viruses, usually takes some decades to build up. The whole flu virus globally mutates each year, making it far more effective against us and more dangerous (in one sense this seems particularly creative and clever of the virus, but it's simply the result of a particularly unstable [RNA-based], fluid viral genome, with older versions losing their foothold because of mutations and widespread immunity in the infected population following the infection). Newer versions are able to infect us again, as if we've been exposed to a different virus.

There are two complementary views of this viral gate-crashing of an organism. From one perspective, the virus is simply exploiting its newfound information about a decent potential home. A slight modification of this view is from the position of a single gene and the evolutionary pressures that it faces. Say this gene does indeed capture something useful about the world. It might nevertheless be amid a bunch of bad ideas, living on a sinking ship of a rather inadequate virus species. But if the virus becomes a new composite life-form, say, by turning into a portion of the DNA code within a bacterium, then this act might just save the gene from drowning, and the good idea is preserved.

But, from another viewpoint, the host organism may not strive *too hard* to repel this ready-made set of potentially useful genetic ideas, since one or two may spell survival in an otherwise catastrophic environment. With these interspecies packets of DNA information, there comes the enticing possibility of a substantial innovation for the organism, of a novel perspective in a difficult, dynamic world.

There is in fact increasing evidence that this chopping and changing of DNA is an incredibly common process, vital for evolutionary success, and present in all species, including humans. And while other interspecies gene-swapping is a powerful, though admittedly dangerous, way of absorbing new genetic ideas, by far the greatest source of evolutionary innovation is the virus.

Although viral injections of DNA code are extremely common in bacteria, the human genome is also positively littered with viral material—as much as 50 percent of our genome consists of scars from ancient viruses burning their way into our code. But it's also clear that some of these invasions have dramatically helped us, either by transporting useful results from other species or by shaking up our own code to give the possibility of new traits: For instance, one such viral donation of DNA is thought to have been responsible for the creation of the placenta in early mammals.

This common genetic intermixing, either between similar species or resulting from viral invasion, raises the speculative suggestion that one reason the DNA code of life is so utterly universal is that it facilitates the injection of chunks of novel genetic ideas from diverse sources. If this is true, then even if it can so often appear to be marked by such a cruel sequence of violent battles in the external world, the whole biological realm can also be seen in part as a strangely collaborative process to optimize collections of internal, DNA-based ideas.

COOPERATION AND DIFFERENT LEVELS OF INFORMATION PROCESSING

In science, we know that the universe is not built on a string of an infinite number of unrelated facts. Scientists instead strive to detect the underlying patterns—the relationships between atoms of information and the overall informational structure. Interrelationships are universal and highly layered. For instance, quarks combine to make subatomic particles, which group together in the form of atoms, that bond with others in a molecule, which link up to build a protein, which interacts with other proteins in the mechanics of a living cell, which has its specific role to play inside the organ of a system of a human, who is an employee in a department of the regional building of a national sector of a global firm with branches all over a planet orbiting a star that is part of a galaxy.

Humans have an unrivaled intellect with which to detect, reflect, and amplify information structures. But if information processing really is a deep river surging through all life, it's natural to assume that one of its major veins will be the representation of structure and levels of meaning in the biology of the cell.

The celebrated evolutionary biologist Richard Dawkins has made famous the idea of "selfish genes," whereby the prime locus of evolution is the genes that travel through time, passing via reproduction from one host to the next. Hosts are simply organisms, though in Dawkins' language they are termed "survival machines." Organisms are relegated to being the mere carriers of those collections of genes, each of which evolution has molded to be a supreme, selfish survivor across the generations. There is no doubt that the concept of a selfish gene is a powerful one, backed up by considerable evidence.

In many genomes, though, one needs to go a level below that of genes to find the true winners (Dawkins' definition of "gene" is different from the standard one, and he would class selfish, noncoding DNA as genes, too). Smaller sequences of DNA, not coding for proteins—and therefore not classed as genes in the standard definition—sometimes nicknamed selfish DNA, can hide quite happily in a genome and even replicate like crazy. For instance, there are thought to be up to a million copies of the Alu sequence in each human, constituting a staggering 10 percent of our genome, even though this sequence doesn't actually code for any protein. No organism is likely to ever support such utterly prolific, selfish reproduction of a gene. But these tiny sequences of DNA have found winning ways to live and breed within the world of the chromosome, remaining largely invisible to the rest of the cell, let alone the outside world.

Evolution isn't confined just to genes, though, or to their baby brothers, these small DNA strands. Darwin's theory of natural selection was based on the organism as a whole, and this level still, in some ways, feels like the most useful one to focus on when discussing evolutionary pressures, since it encompasses not just the sum of each individual gene within a creature's genome, but also any genetic ideas that emerge over and above this simple sum, via the complex ways that genes and their effects can interact. For instance, the creation of the human brain involves thousands of genes in an incredibly sophisticated, intricate collaborative enterprise.

Occasionally there may well be more purely selfish genes that survive in ways that inhibit the survival of the host. Humans are prone to so many genetic disorders that it's amazing we survive at all. Actually, if some selfish gene were *really* disadvantageous to its host, at least early in life, the host would die before reproducing, and the gene would be lost. Over many generations, the gene, on average, that promotes the well-being of the host is more likely to hitch a successful ride through the generations. So there is a pressure for genes to be selfish, but *in an enlightened way*, to collaborate, coordinate, and play their small part in ensuring that their host, this "survival machine," flourishes.

Admittedly, it is somewhat a matter of perspective, but I believe something is missing from the view of evolution as individual genes striving for immortality via this hopping maneuver between organisms. Instead, a more parsimonious way to describe evolution may be that it is an active competition of ideas for survival, with those concepts more accurately capturing relevant details of the world being more likely to persist. This is a deliberate overgeneralization—such a definition would include fields such as the scientific enterprise and capitalism. Crucially, though, for biological evolution, although this perspective shares the assertion with the selfish-gene position that organisms may just be stepping stones for something more central to persist through time, this ideas-centric definition places no limits on the domain in which evolution works, potentially applying to any level at which ideas compete.

Another, related difference is that a selfish-gene view would prefer a gene to resist any form of change, whereas the ideas-centric view would assume that a gene may welcome change in its own code, if this is related to a better idea, including one that is represented collectively by a set of genes. Dawkins has argued that manipulations in mutation rates, say, are the selfish imposition of a single mutation-causing gene on the unwanted identity changes of all the others. In heavy contrast, a perspective of evolution in terms of the primacy of blind ideas would suggest that all genes may appreciate such erosion of their identity, when it's clear that their ideas aren't fitting with the world, and that such a manipulation of mutation rates, in such a pointed way, and as a reaction to heavy stresses, is a solid collective strategy to increase the chances of finding saving innovations.

This potential for atoms of ideas to be built up to higher concepts is a crucial feature of evolution. If my mind could not ever combine basic features of the world into objects or categories, if all memories for me could only be incredibly simple, single ideas, such as "black" or "a dot," instead of "computer" or "fruit," then my understanding of the world would utterly disintegrate. Similarly, many ideas might well apply at the simple level of the gene, but other, more sophisticated, and possibly far more useful, blind beliefs might require the complex interaction of hundreds of genes; at higher levels, certain concepts might even exist only by the way that many organisms behave as an ensemble. Intriguingly, there may well be an evolutionary pressure toward these more advanced ideas, made up of lower-level components, since these compound concepts would tend to be more accurate, intelligent, and powerful.

Indeed, the most common level for useful ideas to emerge is almost certainly that of combinations of genes. In analogous fashion, massive cooperation is an integral aspect of how the brain processes information: A single neuron represents only a tiny portion of one memory, say the face of my daughter, but it will also represent tiny portions of thousands of other memories, too. My memory of my daughter's face is not carried by a single neuron, but is the emergent property of the interactions of thousands. Likewise, a complex multicellular organism will have tens of thousands of genes or more, and any one gene might have multiple functions and only play a small part in creating any one trait. One very telling, multilayered example of this point is that 20,000 genes, 80 percent of the entire genome, are required to create your brain and to support its proper functioning so that your consciousness will flourish.

Even before true life began, chemical components within a proto-creature would almost certainly have combined to make a conglomerate that carried an idea that was better than the sum of its parts. There was no design or magic to this blind insight—just the random combination of physical building blocks, and an evolutionary pressure to favor any possibilities that were superior at remaining stable and making copies. Within life, there is enormous scope for the emergence of complex ideas formed from groups of genes, owing to the millions of generations and billions of interacting genetic ideas that each single organism represents.

Some forms of bacteria, for instance, combine forces in aggregates, generating far more successful defenses than is possible alone, and within this collaborative structure different bacterial cells can even take on different roles, so that they closely resemble a multicellular organism. One stage further, true multicellular organisms use extensive division of labor, with many different cell types each playing their small part in keeping the organism alive and able to reproduce.

Although there are famous cases of close, intelligent collaboration among animals, such as social insects, there is increasing evidence that plants communicate and collaborate as well. For instance, reminiscent of wealthy older people becoming philanthropists, Douglas firs have been known to share soil resources with saplings of the same species (not just direct progeny) via underground fungal networks. And when tomato plants are attacked, they release both airborne and underground chemicals that neighboring plants can read in order to raise their own defenses.

It also appears, intriguingly, that some ecosystems that are particularly robust to change manage to self-organize structural properties. In Niger, for example, there are sections of dense vegetation alternating with barren regions, forming patterns resembling a tiger's stripes, so that ecosystems can continue to maintain some vegetation even when the average resources would otherwise be too low to support it.

This possible, high-level, "intelligent" information processing still has its foundations within the DNA that is the ultimate source of such behavior. But the connection between this top-level, ecosystem-based idea and a single section of DNA is increasingly remote: This concept at the apex of an informational hierarchy is supported by the knitting together of ideas on each of the multiple levels below (ecosystem above collections of organisms above physical characteristics of those organisms above the interaction of genes, and so on). Perhaps this relationship is just as remote as that between a conscious thought and the activity of a single neuron.

And although talking about evolution beyond the realm of the organism is controversial, wherever there are competing ideas between information carriers capable of change, something resembling evolution in all but name may well occur.

It's possible that some particularly complex ideas can really only be supported by a group of organisms, and that if an idea helps keep them all alive, then, again, evolution could, in principle, step in to favor this information chunk. Although the transmission of that information down the generations still involves genes, the point of evolutionary pressure essentially resides at the level of the concept, in this case the group of creatures, as if they were a single system. In similar fashion, we wouldn't claim that the stock market rose 1 percent today because of the physical laws that govern how fundamental particles interact, even though the stock market wouldn't exist without such particles.

So in this way, from the level of short sections of DNA all the way up to ecosystems and beyond, evolutionary pressures could in principle weed out those ideas incongruent with survival, while favoring any concepts that capture something accurate and crucial about the world. And one part of this process may well be the encouragement of the complex combination of ideas at the lower level to form more enlightened blind concepts a level above.

Genius Cells

So far I've only discussed information management within DNA. But if the other organisms around you are performing the same genetic informational tricks, how do you inch ahead in the evolutionary arms race? One potential way is to start storing and changing information using other tools within a cell, to build additional layers, not just in terms of the domain and structure of ideas, but also in their computation.

In almost every realm imaginable, science has been revolutionized by computers. For a couple of days last week, I analyzed a large fMRI dataset— or rather my computer did, since there were well over 3 billion calculations needed to reach the results. Computers are unrivaled tools for the scientist, helping enormously in both the collection and analysis of information.

Similarly, if an organism won the random mutation lottery and got its hands on a better biological form of computation, the rewards would be enormous.

So far, DNA-based ideas can only get updated by evolution; in other words, by generations of organisms passing by, and those with genes— or collections of genes—that are able to persist over time being selected over those that cannot. That is in some ways a painfully inefficient way to learn something about the environment, with many millions of life-forms extinguishing before the lesson is fully learned. A far better approach would be to gather relevant knowledge about the surroundings *within the lifetime of the organism.*

This sounds like the realm of animals, but in fact many single-celled organisms, including bacteria, process information in this dynamic way.

The main mechanism available is the proteins that genes create. Some proteins can interact with each other to follow the rules of logic in order to perform rudimentary calculations. Other proteins help sense details of the environment, while some even turn back to the DNA that created them, and turn on and off various genes, thus changing the production levels of other proteins. These new layers of information communication allow for a very complex cascade of activity, and surprisingly intelligent forms of learning and ideas.

One collective example of the generation of a complex concept are Hox genes, which control the location and number of developing limbs in animal embryos by deciding whether other genes are activated. Some of these controlled genes, one step down in the hierarchy, themselves regulate the activity of other sets of genes.* This is highly reminiscent of many aspects of human life, such as the network of staff in a large corporation, or the many layers of categorization we mentally learn (for instance, my laptop is a kind of computer, which is a type of electronic device, which belongs to the set of machines, all of which are a form of tool, which are inanimate objects, and so on). Most complex systems benefit hugely from a hierarchy of knowledge and management, single cells included.

Far more impressive, though, is the facility for learning that microbes can demonstrate, usually via these protein-based computations. Bacteria, for instance, can communicate with each other using chemicals to indicate a lack of food, and thus each bacterium will spread out in a region to maximize consumption of what little food is available.

---

*The human genome includes around 23,000 genes—making it far smaller than that of many other organisms and surprisingly minuscule compared to what you might expect for an organism that contains the most complex organ on the planet—the human brain. However, by using many clever techniques—one gene coding for many proteins, hierarchies of controlling genes, and so on—we make the most of our meager genetic lot. A better index of organism complexity than number of genes is probably the range of proteins that an organism makes—and on those terms we are indeed heavy hitters.

Protozoa and bacteria even use rudimentary forms of learning and memory when faced with different types of food or possible threats. For instance, if gut bacteria find some appropriate food, they will ready themselves to digest related food that's likely to be nearby, as if making a kind of prediction, but will stop this behavior if they do not find it soon enough.

INTERNAL EVOLUTION

So evolution favors an accurate internal picture of the world via effective learning. But there are important limiting factors to this process. For one thing, as your internal model of the environment increases in accuracy, more energy is required to maintain this growing set of knowledge, and you become more vulnerable when food supplies fall short. And generating an increasingly large set of ideas requires an increasingly complex organism, so your reproductive rate slows down. However accurate your set of internal beliefs about the world are *right now*, the world can change catastrophically and instantaneously, and if you are sluggish at making copies, there's little chance the critical DNA component of your ideas (if you have others, such as mental memories) will be able to update fast enough to track the changes, making extinction far more likely. Finally, if you have to become a larger, more complex organism to store all these extra ideas, then your bulkier biological machinery is also more likely to break down.

Bacteria hit that sweet spot of just enough complexity, but without it being an undue burden on survival. Consequently, they are capable of surprisingly clever information processing, but they are otherwise simple and small enough to replicate quickly and efficiently. They are the most successful type of creature on the planet by any yardstick you'd care to use: by numbers, because there are a staggering 10^{30} of them; by diversity, because they live not just on all continents and in all climates of the world, but also in acid, in radioactive waste, and deep in the earth's crust; and even by longevity, as bacteria have been known to spring back to life after lying dormant for tens of thousands or even millions of years. Bacteria existed in vast quantities across the earth billions of years before animals turned up, and it's very likely that they will still be around long after humans have perished. Based on this evidence, it seems highly plausible that there was an active trend, via evolution, from the origins of life onward, to favor those creatures that could process information most dynamically and accurately—but only up to the complexity of bacteria.

So why do animals exist in the first place? Part of the explanation is that they arose and succeeded by chance: Given sufficient evolutionary probing over sufficient time, with the right conditions, ever-increasing possible niches of survival will be explored, or, in other words, ever-increasing sets of biological ideas will be entertained. Animals are just one random set of strategies for survival. Of course, humans are a fascinating, wondrous example of what an organism can become, with our rich consciousness and deep intellect, but evolutionary success is a different matter. Having a brain such as ours, for instance, seems to lead to runaway processes that endanger our own existence—excessive CO_2 emissions being one catastrophic example of a set of damaging products of our great collective consciousness.

Leaving these caveats aside, I now want to explore the details of this niche that animals exploit. Modern bacteria can form and adapt ideas immediately by encoding information not just in DNA, but further afield within the cell, mainly by using protein to represent additional ideas, or, even more powerfully, by building many computational links between DNA and proteins. Although ingenious, this system is also terribly limited, since only incredibly rudimentary information can be learned, moment to moment. So what else can be co-opted to manage even more information, if better computational power is a potential evolutionary niche that is to be exploited? With bacteria combining to represent ideas about food as one primitive example, the next logical step is to move beyond the confines of the cell wall.

We are now in the realm of multicellular organisms, with cells specialized for specific functions within the organism. Bundles of nerve cells making a brain are one route nature took, with the evolutionary "hypothesis" that learning and storing even more information on the fly would compensate somewhat for the greater investment of time and resources required to maintain this organ.

Some basic change in the world may take non-animals—including the cleverest bacteria—generations to encode via natural selection and DNA. But even the simplest of animals can, strikingly, learn a wide range of lessons from the environment over just a few seconds. Other more complex features of the world, assimilated easily by animals, may never be captured by DNA alone. In this way, a threat that would have destroyed a non-animal organism, or even a whole non-animal species, because of its limited capacity to process information, might not even harm an animal.

If you view evolution essentially as the competition between ideas, with the best ones eventually claiming victory, then animals are in a sense clamping on an additional, internalized version of evolution in order to enhance their chances for survival.

Thus all life undergoes genetic hypothesis-testing via evolution. The feedback about whether your concepts are right or wrong usually comes from the environment directly, which selects those concepts for persistence across the generations, and the bad concepts for death. If you happen to be a sophisticated type of bacteria, then a small but vital component of the feedback you receive can come from the intermediate steps of proteins, which help sense and adapt to very crude features about the world on the fly.

But for animals there is an additional buffer to process an important subset of beliefs that really matter for survival and reproduction. Feedback still comes from the environment, but much of that feedback need not affect DNA at all, since it can merely change the ideas stored in brain cells. And, in combination with movement, animals can now actually interact in pointed ways with the world in order to test beliefs very actively. The number of possible ideas an animal can entertain in a lifetime is effectively infinite, especially since wrong ideas no longer risk death.

Moreover, the more mentally complex the animal, the more elaborate its internal model of the world is. Thus, much of the environmental feedback that used to be required to change a belief, whether genetically or neurally stored, can now occur entirely within the complex, structured, internal environment of the animal's brain.

Animals with particularly complex brains could even test many competing ideas without moving a muscle. For instance, in the middle of the night, unable to turn off my consciousness sufficiently to fall asleep, because I'm obsessively thinking about consciousness science, I feel a sharp hunger pang and conclude that the best course of action is to obtain a very large bag of cashew nuts. I initially decide to visit the kitchen, but then recall that a now rather irritating spring-cleaning the previous day cleared out most of the food. I then imagine the usual situation of going to the supermarket, but realize that my standard one is closed after 9 p.m. So I either could go to the

24-hour supermarket, which is a 15-minute drive away, or a gas station a kilometer away. I can work out the optimal way to obtain a much-needed, intensely fattening snack, potentially from miles away, without ever leaving my bed, which is in some ways incredible.

This illustrates that evolution has begotten a form of internal evolution, and this internal evolution becomes ever more apparent the more intelligent the animals are, to the extent that we humans have brains that very much behave like internal evolutionary worlds.* We represent the world so fully, so accurately, that we can play out scenarios in our heads and explore a large range of options—all while hardly expending any physical effort. Such experimentation is now as safe as it's possible to be—we don't risk anything whatsoever by searching through the options in the mental realm—not survival by genetically betting on a loser, not even physical damage by learning painfully from our mistakes. This seems a universe away from the proto-life "ideas" with which we began this chapter, but it's not. It is merely a sequence of connected evolutionary steps, all based on the theory that effective information processing naturally confers an advantage.

THE COMPUTATIONAL LANDSCAPE OF A BRAIN

Given the last few paragraphs, I should emphasize again that animals are not necessarily superior to other organisms in terms of evolutionary success. An oak tree, for instance, with its working hypotheses that physical toughness is highly protective and that the sun is a plentiful source of energy, may be just as long-lived and populous a species as a mouse. It's just that animals have a fascinating, powerfully pointed set of advantages converging on complex information processing, and the overall genetic assumption of these organisms is that these few profoundly superior traits outweigh the many limitations.

---

*In actual fact, most animals have a second active form of internal evolution, via their immune systems. Because the range of parasites we face is vast, and their frequency is incredibly high, the risk of death from these invaders is very real. Therefore, we need an immune system that can cope with virtually any eventuality. The way our immune systems can "learn" to combat invaders is very similar to standard natural selection—or brain processes. The immune system creatively generates many alternative possibilities, has those alternatives interact with the pathogens, and then, when some particular possibility finds a match (i.e., it has learned something), the hypothesis that such enemies are around is classed as well founded, and this successful antibody effectively breeds, so that it becomes a prominent feature of the immune system itself.

The simplest benefit an animal gains from a nervous system is the regulation of its basic states (homeostasis). This computer in the head is able, in almost every animal, to help control important biological features. By monitoring the animal's internal temperature, for example, and initiating any needed response when the animal becomes too hot or too cold, it can keep this temperature within the optimal range, acting just like a thermostat. Even in cold-blooded animals, where temperature is regulated entirely by the ambient heat, this can be achieved by directing the animal to move into a hotter, less shaded region if it is too cold. This single regulatory process provides a powerful advantage for many animals over non-animals—namely, that they can exist in a wide variety of locations, and may not need to shut down for winter. When you have a computer around, you can also fine-tune many other features, such as how much energy (glucose) or water there is in the blood, the concentration of salt, and so on. In each case, there is a monitoring system and, if necessary, a chemical messenger (a hormone, usually) that cause a change to correct any form of imbalance.

But this internal regulation is only a tiny portion of what an animal brain does. The main purpose of a brain is to sense the outside world and move around based on this data. Retrieving accurate external information confers a significant potential survival advantage. Although some bacteria, with incredibly crude sensory skills, can detect via protein switches when food is scarce, which is indeed very clever, such an organism would look remarkably stupid if food was actually at the center of its world, but undetectable by it just because it was hidden by a chemical barrier. An animal with multiple senses might be able to look for food, immediately see exactly where it was, what it was, whether other animals were feasting on it, smell how energy rich it was, and hear if there were any predators nearby waiting to pounce as soon as the animal tucked in.

We take our senses so much for granted, and rarely perceive them for what they really are. Our senses are nothing more than conduits to pick up physical information about the environment: a small portion of the electro-magnetic spectrum for vision, for instance, which almost everything around us reflects or even emits; the compression waves of air or water for sound, which many moving things generate; and chemical offshoots of interesting objects for smell. Our different senses feel utterly distinct to us, but there is an important bottom line here: *It's all just information.*

One demonstration of this comes from experiments on ferrets. If you rewire the ferret visual pathway from birth, so that instead of going to the visual cortex it ends up in the auditory cortex, then the auditory cortex, which should be specialized for hearing, ends up doing a pretty good job of processing vision, allowing the ferret to see. This auditory region will even take on characteristics (such as having neurons specialized for representing the angle of an object) that the visual cortex would otherwise process. It isn't quite as good as if the visual cortex were doing the job, but it is quite functional. Another striking example comes from people blind from birth, who, when reading Braille, process the words mainly in the visual regions of their brain. Given that the visual regions have no sight-based input with which to work, this portion of cortex has adapted to take on the Braille-reading task instead. These sorts of examples show that all sensory processing is just information to the brain, and that almost any brain region can process any type of information, even if it was originally earmarked in development for a specific type of data.

Importantly, especially in more complex nervous systems, our perception of the world is far from being a mere copy of the physical information hitting our senses. Instead, an active, ever-changing, yet unconscious statistical machine is in force, transforming the basic information we receive into a detailed model of the immediate world, including how it is likely to change in the near future, and what in the environment is particularly relevant to us.

But there's no point just filling up your sensory bank with information, however well processed, if you're just going to hoard it, never using that hard-earned knowledge for useful purchases. What's needed is a way to tie information to behavior—in other words, to move. In simple animals, this is all too clear, as the connection between what they perceive and how they act is usually direct and almost immediate: A worm, sensing food, will go toward it, or sensing some threat, will move away. But the more sophisticated an animal, the more processing takes place in the gap between senses and movement.

For humans, the brain's main evolutionary purpose—moving the body— is rather obscured by just how much happens inside; the link between what we sense and how we behave is a long, fragile tangle of thought. The processing and the calculation of what action is best out of the millions available to our finely tuned bodies and remarkably accurate world picture has taken center stage. Nevertheless, it's useful to entertain the interesting perspective that, essentially, even a human brain is there primarily to move the body around in the most useful ways.

The first mechanism for movement is instinctive behavior, a tool that all animals possess. An instinct is a genetically determined brain program to marry some sensory input with some prescribed response—designed, as always, to maximize the survival and reproduction of the animal. Here the genes are basically exploiting the ability of the brain to store and act upon information too complex to be executed only in a direct genetic form. For instance, if I unwittingly touch a delicious pie that has just been removed from the oven, before I know what has happened, I find my hand darting away from the potentially burning source. It takes a moment or two for my consciousness to catch up and realize that I nearly burned myself. The primitive regions of my brain sensed the heat and programmed the response before my higher cortical regions were even informed, either of the heat or the arm twitch. It all looks very simple and natural, especially when consciousness appears to be merely a spectator at the event, but you still need some surprisingly complex neuronal processing to make this important reflex occur—you need to know in which direction to send the arm, which muscles to trigger, by how much, and so on.

Matching responses to fixed sensory input is all very well, but on its own, it doesn't get you all that far. In one episode of *The Simpsons*, Lisa's original school science fair exhibit is destroyed by her mischievous brother Bart. In revenge, she decides that Bart himself will be an unwitting component of her new exhibit, so she carries out a series of tests to determine whether Bart or a hamster is smarter. In one experiment, the hamster starts nibbling at a piece of food, but receives a mild shock. It immediately learns never to touch the food again. Cut to Bart, who discovers a cupcake with electrodes attached and a sign in front boldly warning, "DO NOT TOUCH." Chucking the sign behind him nonchalantly, he reaches for the cupcake, and his hand darts away as he receives a sharp electric shock. But this just makes him angry, so he reaches again. Surprise surprise, he's shocked again and his hand darts back. He continues repeatedly to reach for the food and repeatedly shocks himself. In the real world, I don't think Bart would have made it to the age of ten. But what he—and the hamster—clearly illustrate is that it's all very well having instincts, but without even the most basic kind of learning, they don't get you very far.

Even the simplest form of learning is usually surprisingly powerful. If an animal tries to eat a food source that is toxic, it will quickly not only link this input with the result, but also avoid anything similar in the future. This is an incredible feat. The animal has managed to link in memory a useful, abstract copy of the object with its effect (more intelligent animals will add to this mix the relative intensity of the effect—for instance, chocolate is more tasty than spinach). What is more amazing is how few brain cells you need to learn in relatively sophisticated ways like this. Returning to the humble nematode worm (specifically *C. elegans*): This tiny worm, all of 1 mm long, has exactly 302 neurons. Nevertheless, it can learn to connect an arbitrary, neutral smell with a nearby food source and approach the smell whenever it is presented, presumably in the assumption that food is soon to follow. It can learn to stop moving away if some initially potentially dangerous stimulus repeatedly shows itself to be harmless, thus scrubbing out a previously firm, important belief. The nematode worm even shows crude seeds of socialization, with some strains only stopping to eat when in a group.

When you move up in sophistication to the fruit fly, another commonly studied animal, you can create a surprisingly brainy biological machine with only about 200,000 neurons. As well as learning by simple association, like nematode worms, fruit flies sleep, have short- and long-term memories, and even have a primitive analogue of attention—all inside a brain the size of a poppy seed.

Learning is all very well, but what should I try to learn when surrounded by infinite potential facts? And *why should* I bother to learn to avoid the pie until it cools down? *Why should* the worm bother avoiding some type of toxic food? Why not learn instead that the wind makes that leaf there on the right bob up and down a little faster than the one on the left? The crucial answer is that animals are constrained by a value system—the representation of what's good or bad, pleasant or painful. Animal behavior is closely governed by this system. And this mechanism, via evolution, has been honed to closely map that which is beneficial or detrimental to survival and reproduction.

This value system labels any remotely relevant stimulus according to whether it will aid or imperil the animal and how great the danger or benefit is. Simple animals will enshrine this in their movements—they will approach what is good (such as food or sex) and escape from what is bad (such as predators). In many animals, the speed and level of permanence of learning is also related to just how beneficial or dangerous the source is. For instance, a dog may almost immediately learn that when its owner starts screaming at it, a kick is sure to follow, but it might take many repeats of its owner calling out "new water" for it to realize that there is something to drink again, since water is in plentiful supply, and its thirst is rarely life-threatening.

Though a simple animal may learn what is good or bad about the environment, this doesn't explain *in what way* it is good or bad. *C. elegans* will back up if it smells something paired with a toxic food source in the same way that it backs up if it senses a vibration—there in fact is little distinction in its brain between the two events. This is where emotions extend this value system. Emotions put meat on the bones of what is beneficial or harmful. The three main primitive emotions are fear, disgust, and anger. If we're afraid of some smell, we sprint away—or freeze in cover—while also being far more alert to the danger, ready to notice its slightest detail, and actively prepared to escape again if necessary. If we find that a smell evokes memories of a disgusting meal, it's just plain silly to sprint away through the forest as fast as we can. Instead, we will merely slowly back off and look for something else to eat. So basic emotions can shape our behavior in far more sophisticated, prepackaged ways, in relation to different categories of threat, than a crude value system that only ever has two labels: good or bad.

Some psychologists have suggested that we recognize our own emotions wholly by what we pick up about our body states. When we're angry, they suggest, we're actually just noticing that our heart rates have increased, our fists are clenching, and so on. Again, emotions are largely a signal to move, to change the environment to maximize our survival and reproduction within it.

Of course, humans, in contrast to simple animals, have a large range of different, more complex emotions, such as jealousy, schadenfreude, or a sense of injustice. The variety of our feelings has increased dramatically compared with many other mammals because our evolutionary heritage is that of a large, complex, hierarchical society, which places many more demands on managing social politics and a diverse group of friends. Some emotions may destroy our lives, such as the obsessive love we may feel for an unavailable woman, or the addictive rush of gambling. Nevertheless, all the feelings we experience have a clear evolutionary foundation. Our brains perceive features of the environment to be good or bad for us, even if the computation can go askew at times due to the complexities of our lives. And many complex, seemingly subtle emotions are combinations of simpler ones, each with a clear evolutionary purpose. Jealousy, for instance, is a deep desire for some object, such as a mate, and primitive anger at the threat to our possessing this object.

In fact, the more we study our chimpanzee cousins, the more emotions we find that we seem to share with them, and the more apparent it becomes that our own large set of sophisticated emotions have an evolutionary underpinning. The latest research additions, for instance, include findings on the existence in chimpanzees of a moral sense, and possibly even a degree of wonder. If a chimp encounters a waterfall, it will sometimes display in front of it, touch it, and stare at it for prolonged periods of time—even if no other chimps are around (thus helping to rule out the possibility of an alpha male showing off to assert his authority over his group). If there is a thunderstorm, both males and females have been known to play out a kind of dance. Some primatologists have speculated that this is because our simian cousins can occasionally show intense curiosity, even a kind of reverence, for dramatic displays of nature.

VAST INTERNAL WORLDS

I'm not, for the moment, assuming any awareness in any animal aside from ourselves. But at the same time, I firmly believe not only that our emotional repertoire closely links us with chimp minds, but also that there is a continuous thread running from a husband's bursts of jealousy all the way down to the battles between Alice, Beth, and Claire earlier in this chapter.

Although all life is predicated upon capturing useful ideas about the world, there is a remarkably common tendency for information stored on one level to combine to create a richer concept at a higher level. In some cases, further layers can be constructed on these foundations of foundations, and so on, until an efficient yet towering edifice is created. It leads to bacteria using control genes, or finding computational switches for rudimentary learning, or combining forces so that all can optimize some food source. It leads to shoals of thousands of fish collectively twisting elegantly, easily away from a predator, even though each fish alone wouldn't stand a chance. It leads to millions of ants together developing highly intelligent behavior, partly via simple chemical signaling. It also leads to evolution creating a value system as a shortcut for hypotheses about what helps or harms an animal, then building simple emotions on top of this, and then stacking complex emotions on top of the simple ones. And it leads to humans developing categories, plans, language, and ingenious strategies to serve, modulate, and enrich our emotions and motivations. This is partly because nature really is highly structured, interconnected, and hierarchical. Within each level, complex structure can be discerned—sometimes so highly ordered that simple mathematical equations can capture almost every detail.

Information in the universe is brimming with patterns. Scientists aim to discover those patterns with sufficient accuracy that they can make staggeringly accurate predictions (how else can a satellite sent on a 500-million-kilometer, half-year journey into space arrive at the precise orbit around Mars that NASA scientists wanted?). But it's not just scientists who benefit. As biological machines with large, complex brains and pressures to innovate screaming at us from every angle, we all have good reason to spot useful patterns rather than just simple facts. The more accurately we represent the structures of the universe, the more control we gain over the environment and ourselves. This is true both in terms of scientists, research, and technology and from an evolutionary point of view. Both have benefited from the invention of incredibly powerful hypothesis testers. In humans, the connection between scientific research and our fundamental intellectual qualities seems so similar as to be trivial, but we came in the first place to these incredible mental faculties precisely because of evolution favoring accurate information processing, almost by definition, ever since the first proto-life creatures emerged in the oceans.

Exploiting patterns isn't limited to animals. You can find patterns everywhere in nature. Many viruses adopt a regular circular shape, partly because the smallest, most efficient genetic instructions are required for such a regular structure. Flowers form highly symmetrical shapes with their petals or seeds—patterns that mathematicians immediately recognize, and that are attractive to bees precisely because of the order the bees perceive in them. But non-animals are severely limited in terms of the regularities they can encode.

Many animals are constantly, almost desperately, looking for patterns. Occasionally, this search can go rather wrong. Burrhus Frederic (B. F.) Skinner, one of the pioneers, along with Ivan Pavlov, of the study of simple forms of learning, found that if he presented food at regular intervals to a pigeon, without any cues preceding the feeding, the bird would nevertheless manufacture some action to pair with the tasty stimulus. One bird carried out a bizarre dance, twisting anticlockwise a few times. Another carried out a repeated head nod. It was as if the bird believed that twisting around would generate food. Skinner pointed out that this behavior looked remarkably like human superstition, which can result in such things as rain dances, or beliefs in astrology. It does seem clear that one component of the extreme popularity of irrational beliefs such as religion or alien abductions is our unerring search for structure and meaning within the constant torrent of information to which we are exposed.

The prevalence of superstitious beliefs in the animal kingdom suggests that some underlying process closely related to it is useful for animals. Imagine that the extent to which an animal searches for useful patterns is like the volume control on a low-grade stereo. Turn it too low, and you can't hear any detail. If the volume is halfway, you can hear quite a bit of the music, but some of the accompanying instruments are hard to pick out. Turn it to the max, and much of what you pick up is distortion; you can hardly hear the main tune. Animals, it seems, like their informational music pretty loud; they want to be able to pick out every interesting detail in the melody, as well as the bass and the accompanying instruments, and achieving this highish volume is worth living with a bit of buzz and distortion. It's better to maximize your chances of exploiting opportunities to learn new, interesting features of the world than to miss these potential insights, even if, occasionally, you make false guesses and latch onto irrelevant behavioral habits. This is highly reminiscent of chaos-inducing strategies that occur on a genetic level, such as mutations, sex, and jumping genes, which may create many fatal novelties, but also some genuinely useful innovations.

But while the innovations in bacteria were impressive, when you have a large biological computer, such as a brain, capable both of storing stable ideas and learning new ones on the fly, the gains in information processing and physical control of the environment are exponentially improved.

This constant searching for tricks and patterns regularly yields solid dividends, to such a degree that we are largely blind to how vast our collection of structured knowledge is and how much it dictates who we are. I'm not alone in being a huge fan of tennis giant Roger Federer. I'm constantly astounded by his accuracy, his power, and the diversity of his shots, which seem so many orders of magnitude above what humans should be capable of. But I take comfort from the fact that he is merely mortal, and spent many thousands of hours obsessing over and practicing his art, starting not long after he was out of diapers. In almost any field, spending thousands of hours devoted to your beloved topic seems an important prerequisite for appearing to be a genius. But, really, we are all super-super-Federers in the way we mentally interact with the world, and the reasons are similar—practice. This time, though, the practice is largely on an evolutionary level. We've had not thousands of hours to hone our skills, but half a billion years.

Most animals have a combination of highly practiced models of the world—some are genetically determined; others are merely primed by genetics; and still others are entirely prescribed by learning and experience. Humans are unusual in the animal kingdom in that although we are born fully formed, with all the parts where they should be, we are helpless for many months—normally having to wait for over a year before we can even walk. Part of the reason for this is somewhat trivial: If we were born with our brains properly developed, we would have to be born with much larger heads, and our mothers would require pelvises so large they would hardly be able to walk. Another reason, though, goes to the heart of what it means to be human. In this seesaw between fast, brittle, uncomplicated instincts and slower, flexible, potentially complex plans, humans definitely weigh heavily on the side of the latter. We begin life embarrassingly ignorant and incapable, but have an enormous capacity to learn and optimize any motor skill and any representation of the world—and that's exactly what we spend our lives doing.

We may not all have Federer's physical poise, but we nevertheless find it trivial to walk about a busy town while chatting to a friend, fluidly moving our limbs and mouths in incredibly coordinated ways, picking up and moving multiple grouped objects simultaneously with undoubted finesse, recognizing obstacles automatically, and so on. All these are utter computational marvels. They are partly the product of hundreds of millions of years of evolution, which have provided us with an immensely powerful biological computer, honed in just the right way for us to understand the world and move within it. But the supremely sophisticated ways in which we navigate through our daily lives are also partly a result of the vast amount of learning we've accomplished by the time we reach adulthood.

In many surprising realms, our apparently effortless feats of thought and motion make those of the most cutting-edge robots look utterly idiotic. Despite many years of research on the part of their creators, any robot that is designed not to perform perfectly prescribed, repetitive movements, but instead to learn about perception, walking, and the meaning of objects, tends even now to fail at these tasks in a comical, catastrophic way. The robot ends up resembling some combination of a heavily alcoholic squirrel and a newborn child. This is despite the stupendously powerful computers we now have at our disposal and much genuine progress in artificial intelligence and engineering. We might not realize it, but almost everything we do so automatically and flawlessly is in fact frighteningly complex, requiring a vast infrastructure of carefully sculpted biological algorithms.

The brain as a computer is a fantastically complex statistical engine, constantly refining models of how the world is and will be. Everything feels so easy to us, but that is all an utter illusion, just as Federer's skills seem so deceptively effortless (until you pick up a racket and try to emulate his expertise yourself). In almost every corner of our mental lives, we are applying a particularly powerful form of predictive statistics (known as Bayesian inference), which boils down to tweaking our model of current and future events based on related events from the past.

This combination of frenetic statistical computation and a constant, conscious, roaming search for patterns allows us to aggressively, minutely track thousands of flittering features of the world and discover insights about them to recreate so much of its beautiful structure inside our minds. We are consequently, in some ways, now heavily protected from the cruelties of biological evolution, since so much of the evolutionary battles between ideas now occur in the mental realm. A professional, scientific picture of the universe is one quite natural product of our massive brains. We have tamed and controlled nature to give us plentiful food and protect us from the elements, and we are increasingly able to fight the disease and decay in our bodies through medical advances.

At the same time, our vast capacity to learn—to extract regularities and meaning from that information stream—is intimately related to our awareness, to the rich and deep experiences that constantly punctuate our lives.

3

The Tip of the Iceberg

Unconscious Limits

A Holiday from Awareness

For many years I've played soccer every Wednesday with my Cambridge re-search department. The game is taken extremely seriously—aside from the small exceptions that we never keep score, we swap players around contin-uously if sides become uneven, and we seem to view the pub trip afterward as an event more sacred than the actual match. A few years ago, halfway through a game, I had the ball in front of a particularly large, intimidating member of the opposition. In my head, I was preparing the execution of a complex, deft maneuver: I would elegantly dance around him, along with any other foes in my path, the ball seemingly glued to my feet. I would sprint impressively for the goal, scoring right in the corner with a blistering shot. In reality, on my first body-twist in this foolproof plan, I landed embarrass-ingly on my arse, unable even to claim the conciliatory prize of a foul, since no one had so much as touched me.

On my clumsy journey to the grass, with my backside leading the way, I'd felt two rather worrying clicks in the middle of my right knee. These re-sulted in an abrupt end to the game and a trip to the emergency room, with a possibly torn anterior cruciate ligament.

When the knee failed to recover after a few weeks, the surgeon recom-mended an operation so that he could have a look inside and fix any injuries. The next morning, I was lying in the anteroom next to the operating theater, about to experience general anesthesia for the first time in my life. It's completely understandable that many people are rather anxious at such events. I, on the other hand, felt like a kid in a candy store, albeit a very nerdy, neuroscience-loving kid, with candies made exclusively of brain-altering medications. Earlier that morning, I'd already had an in-depth, chatty discussion with the anesthetist about the wonderful drugs I'd be injected with and which neurotransmitters they'd work on. Now, with the needles in place, I was tremendously excited, supremely curious to discover what effects this concoction of chemicals would have on me.

The first, a muscle relaxant injected into my hand, caused the ceiling to start moving in a gentle, though quite marked, circular path. Then came the main anesthetic, propofol. I'd read research papers about this drug and was particularly intrigued by it (this was before Michael Jackson had died of an overdose of it). I was asked to count down from 10. I might have reached 6 or so. There followed a period where, from my own perspective, I was indistinguishable from being dead, so complete was my absence of consciousness.

If I have a really good night's sleep, on the surface I experience, or rather fail to experience, a similar hole of awareness. But on closer inspection, it is actually very different. There may be 5 or 10 minutes when my thoughts lose coherence, turning into daydreams, followed by random ideas and images peppering my dissolving awareness, before I am finally asleep. When I'm slumbering, there are many dreams that any abrupt jolt into wakefulness might allow me to report. Even when I'm not dreaming, if I'm suddenly woken up, I can often recall a kind of random, mad thought-train. Sleep for me largely feels like a semi-mute background of jumbled, incoherent experiences—albeit ones that I'm constantly forgetting, moment to moment. Even if I'm in the deepest sleep of the night, in some ways this is quite fragile. The cry of my baby daughter, or my wife tapping my arm, switches on my consciousness almost immediately.

But when I was under general anesthesia, it was as if I were seeing the room swim, counting down from 10 one second, and in the next second (actually 60 minutes later) waking up again, though in a quite groggy state. In between, there was absolutely nothing, no sensation, no conscious thought. Not a single noise from the doctors or even a serious trauma to my body could change that unruffled state (which is, of course, the purpose of general anesthesia).

Apparently, some people return from anesthesia to the land of awareness in a fevered panic. I'm not sure what this says about my personality (or beverage preferences), but I regained consciousness feeling pleasantly drunk. I think I immediately asked for some Scotch to be added to my intravenous drip to extend the effect. I then might or might not have made some rather inappropriate jokes to the pretty female nurse, who, luckily for me, must have witnessed all this before, as she responded merely with a few smirks. Five minutes later, to my grudging disappointment, the effects faded, and for the rest of the afternoon I was left with the minor consolation of the morphine euphoria to explore.

In a bizarre way, though, that hour's black hole in my consciousness was one of the most memorable experiences of my life.

In this book, I am closely linking consciousness with information processing. But I am certainly not claiming that bacteria or plants have consciousness, despite their information-processing capabilities. And the simplest of animals have, at the most, a minimal level of consciousness. But what's the difference between us, with our extensive consciousness, and these other creatures? And why does our consciousness fluctuate so much in our lives— for instance, when we fall asleep or undergo general anesthesia? In this chapter, I will use the boundary between human conscious and unconscious processing to help answer such questions.

Unpeeling Evolutionary History in the Brain

General anesthesia is a powerful way to investigate consciousness because it can safely, reversibly, yet comprehensively remove awareness. Therefore, we can use general anesthesia to ask what processes in the brain are missing when there is clearly no consciousness present.

During my own general anesthesia, while my conscious mind had very much left the building, many processes in my brain continued to function in a normal way. I was still able to breathe; maintain my core body temperature, along with a host of other basic functions; and generate surprisingly strong brain waves.

In order to help explain this neural division of labor between conscious and unconscious functions, I first need to make a brief digression and provide an overview of the structure of the brain.

Evolution has created a staggering range of organisms, each with features cleverly honed for its environmental niche. But while evolution is a fantastic creator, adding almost whatever is needed, it is surprisingly lazy at tidying up after itself, at pruning what is no longer required. In the bacterial world, where margins for survival may be razor sharp, things are more efficient. But most animals carry with them a surplus of obsolete features, such as the astronomical quantities of pathological DNA interlopers that sit in every cell in our bodies. But there are also more large-scale examples of detritus we endure. For instance, whenever we get cold, our hair duly stands on end to create a buffer of trapped air around our skins, as if such an action would make any difference to keep us cocooned from the cold—it doesn't (unlike other primates, we simply don't have enough hair to make this automatic response functional).

The human brain is in some ways an even more extreme example of a process that is far more creative than destructive. We effectively have three evolutionary versions of brains in our heads (see Figure 3). Our brains are rather like a city that has existed since ancient times. In Cambridge, for instance, the historic center is squashed inside a fertile bend in the river Cam. This is the core of the city. Here there used to be a castle on a small hill, originally built by William the Conqueror in the eleventh century. The oldest parts of the university, along with old churches and so on, are still there. Over the centuries, housing, university colleges, and research departments have sprung up around this central district. And now, around this second band of somewhat old structures, there are the outer suburbs, with modern housing along with large technology and business parks. Although an unromantic person might be tempted to replace the oldest buildings of the city and the narrow winding roads of the core area with efficient modern streets and buildings, all these ancient places still serve some purpose today. The expense of such renovations simply wouldn't be worth the trouble.

At the center of our brains, too, lies the oldest part, mainly involving the brain stem. This is sometimes called the "reptilian brain" because it is the only region we share with our reptilian ancestors. The brain stem is the gateway between the brain and the body—all sensory signals from the body, from the stroke of a lover on our faces to the pinprick of a needle in our arms, passes through the brain stem to the rest of the brain. In turn, all commands from more sophisticated parts of the brain—for instance, for us to tango or kick a football—are shunted through the brain stem, down into the spinal cord, and then through to the rest of our bodies to make our motor commands seamlessly fulfilled. The brain stem also controls other basic functions, such as our breathing and heart rate.

Because the brain stem is such a critical part of our brains, any damage to it due to stroke, tumor, or some accident is usually extremely serious. Damage can frequently lead to death, or, if not death, then a permanent loss of consciousness. Since all motor commands pass through this tiny part of the brain, damage can also lead to a rare but chilling condition known as "locked-in syndrome," where a patient might be normally conscious, but almost entirely paralyzed.

One sufferer, Jean-Dominique Bauby, produced a very vivid description of this state. At the time of his stroke in 1995 he was the editor of *Elle* magazine in France. His stroke left him in a coma, from which he completely recovered twenty days later, at least mentally. He found upon waking that he was utterly unable to move any part of his body except his eyes—and his head, a little. Staggeringly, by just the blinking of his left eyelid and the calm assistance of a transcriber using an alphabet board to ascertain what letter he wanted to spell next, he wrote a book, *The Diving Bell and the Butterfly*, an impressive memoir about his former life and his subsequent mental prison.

Surrounding the brain stem is the limbic system, which is our evolutionary link to the earliest mammals. This system could almost be called our instinct center. It is here where our sexual orientation and proclivities are determined. Also, our hunger and thirst are partly controlled here. Our temperature is regulated in the limbic system, and our biological clocks tick away via rhythmic neurons in these limbic nuclei. Our primitive emotions, such as rage and fear, are generated here, as are our instinctive reactions to fight or to flee. One interesting hint of the age of this region can be found in the kind of phobias we as humans tend to have. Despite the thousands of deaths every year from car accidents, how many people do you know who are terrified of a Ford pickup? And yet, spider bites almost never cause fatalities anymore, but you almost certainly know someone (maybe even yourself) who just can't bear creepy crawlies. The simple reason for this is that millions of years ago, spiders really were serious threats to life. Again, evolution can sometimes be very lazy about updating.

One special region to mention within this class of intermediate brain structures is the thalamus. This is effectively a relay station, the Grand Central Station of brain regions. It connects virtually all parts of the brain together. With damage to certain parts of it, but with everything else intact, a person will enter a vegetative state and show very little sign of consciousness. The thalamus is an important structure to support our experiences, possibly by allowing information to flow smoothly across every corner of the brain.

One reason that our instinctive fears have not been updated is that many aspects of modern life, such as cars, are just too new for natural selection to have yet found any traction. But another reason concerns our third, most modern brain region, the cortex, which is the outer shell surrounding the rest of the brain (cortex comes from the Latin for "bark"). Only modern mammals have this new neural toy. Evolution doesn't need to update our innate fears because we have such a powerful information-processing mechanism in the cortex. We can learn any new fear we need, and even suppress existing ones, if necessary. The cortex is where our most complex, flexible mental activity resides. It duplicates and can potentially modify and control many of the functions of the other two more primitive portions of our brain, but the cortical version of such functions is invariably far more sophisticated.

The cortex, too, exhibits a surprising degree of redundancy. It comprises four main sections, or "lobes" (see Figure 4), but each lobe has a twin on the other half (or hemisphere) of the brain. The occipital lobes, rather bizarrely located at the back, furthest from the eyes, largely process vision. The temporal lobes, at the bottom middle of the brain, process hearing and some aspects of language (especially in the left hemisphere), but are also where simple visual processing becomes object recognition. This is where most of our long-term memories live—storage of the faces we've seen in our lives, our history and sense of meaning. The parietal lobes, at the top back, help process our sense of space, as well as of touch. As part of their spatial representation, the parietal lobes may help us represent numbers, and contribute to short-term memory, via holding on to more than one item in space. They have been linked to the ability to boost attention to some detail in the world. But, increasingly of late, the back portion of the parietal lobes have more generally been linked with complex thought, such as when we attempt an IQ test.

In some ways, the frontal lobes are the odd ones out. They aren't devoted to any sensory processing (although they are next to our smell center, which is found at the front lower section of the brain). You can remove a large section of the frontal lobes in one hemisphere and not see any obvious impairment in your patient (the exception to this is right at the back of the frontal lobes, where the main motor strips are located, but we'll ignore this part for the moment). After losing a substantial proportion of his right frontal lobe, for instance, a patient will in all likelihood still be able to move normally, will have intact senses, and will suffer little, if any, memory loss. One subtle change was outlined in Chapter 1, with the example of Phineas Gage. He lost the frontmost section of his frontal lobes, and his personality was radically altered as a result. This region, known as the orbitofrontal cortex, is now thought to be a secondary emotion center, supplementing the more primitive emotions of the limbic system, the second-oldest evolutionary band of human neural territory. It is in the orbitofrontal cortex that our most complex emotions are activated, such as how we feel and act in social settings and how we convert levels of risk or reward into decisions to act.

Much of the rest of the frontal lobes is responsible for our most abstract thoughts. This area has most closely been associated with IQ, as well as with virtually any task that is either very complex or novel.

The frontal lobe, especially the mid outer part, known as the lateral prefrontal cortex, is thought most centrally to be involved in conscious processes, probably in concert with the posterior parietal cortex, which shares many of its functions.

Before moving on, I should clarify a couple of details. First, by talking about a reptilian brain I'm not suggesting that all the animals from which we diverged earlier in our common histories are so primitive as to be unable to learn in potentially sophisticated ways. Any animal that we became estranged from hundreds of millions of years ago has had ample evolutionary time to develop advanced neural weapons on its own.

Second, these "three brains" living within us aren't in any way independent. They normally work together, in a highly interconnected way, for the common purpose of keeping us alive and thriving. Each level can take command if necessary, and sometimes there is competition between stages within this hierarchy. For instance, we might find ourselves sprinting for our lives, because our fear center, the amygdala in the limbic system, has swiftly taken command after we spotted a poisonous snake in the jungle. For us to sprint away, this intermediate limbic system is using the modern cortical instructions to hurry away, along with the brain-stem instructions to breathe faster, to inject us with lots of energy in order to help us speed away from the danger. But we could also, later on, employ our modern cortex to return cautiously to the snake. We might suppress the intermediate limbic fear messages and deliberately calm our breathing via cortical messages to our brain stem, all so that we might examine this interesting serpent from a safe distance.

As most of the basic functions of our reptilian brains carry on unabated when we lose consciousness, we can safely say that whatever awareness is for, it isn't concerned with those basic internal processes, like breathing, that are vital to our survival. Likewise, consciousness probably has little to do with the brain stem.

However, our two remaining brain layers, our limbic brain and our cortices, enter a special noncommunicative state during anesthesia, making consciousness impossible.

Unconscious Neurons Marching in Step

Our best guess as to the mechanism of most anesthetic agents is that they increase the production of a neurotransmitter, called gamma-amino butyric acid (GABA), that acts to dampen neuronal activity throughout the cortex. In this subdued state, the firing of our neurons becomes more harmonized, and less differentiated, than usual: Strong global brain spikes pulse through the cortex a few times a second in a slow, strong rhythm (known as a delta rhythm). At first blush, it might seem puzzling, even paradoxical, that as our consciousness dissolves, our brain activity becomes in some sense stronger and more rhythmic. But this mystery disappears if we bear in mind the prime, ongoing context of our brains—that they are first and foremost an information-processing device. To explain how information processing relates to brain rhythms, I need to make a small digression to explain just what is being detected when we talk of brain rhythms.

Brain rhythms are regularly mentioned in the press—beta waves for attention, theta for meditation, and so forth. But what does it actually mean to have such a brain pattern? In order to detect these hidden brain waves, you need to use electroencephalography (EEG). This technique involves attaching an array of electrodes across the scalp. Each of these electrodes detects the combined local electrical activity emitted by millions of neurons.

Imagine neuronal activity as a haphazard mix of vacation-goers. These tourists (or neurons) are on a large, boisterous cruise, where, although initially strangers, many are now friendly. On a day trip from the cruise, they are now scrambling in every direction across the rolling hills of a national park near the coast, each chatting briefly with anyone who passes by—for instance, about an accident they witnessed on the road below, or some interesting ancient stone circle on the horizon.

Now, imagine a similar number of people a week later, but this time they are in an army, being ordered by a sergeant-major to march up and down the hills in unison. A distant witness in a passing plane (EEG electrodes) would feel that the soldiers were a more impressive, potent group, and might even imagine there were more soldiers here than tourists in the neighboring hills a week earlier (which there weren't). It looks from the air as if a great swath of khaki green rises up and drops down the countryside in a slow but steady and powerful rhythm. In contrast, the tourists were so spread out as to make it difficult to count them, with far fewer at any one point on the hill summits than now.

But even though the soldiers are more orderly than the tourists, they are less interesting, and also less curious. Unlike the tourists, they don't notice the road, or the stone circle, or really anything besides the gait of the soldier directly in front of them. They are effectively a single group. What's more, anyone caught chatting to his neighbors, trying to liven up the dull walk with a bit of gossip, will be severely admonished by the sergeant-major, and the chatter will die down fast. This all severely limits the power of the soldiers (or neurons) to grab and pass on any useful details about the surroundings.

So, being heavily unconscious, such as under general anesthesia, is associated with a more orderly, slow, thumping march of neuronal activity, which is capable of carrying only minimal information around the brain. These neurons may really struggle to broadcast anything further afield than their immediate neuronal friends; they aren't receiving anything very original to pass on anyway, as so many neurons are singing the same tune.

And here is an important initial clue as to why we can be conscious, while bacteria and plants cannot: Consciousness occurs when there is an active transfer and intermingling of information across much of the neural land-scape. In contrast, any information processing in bacteria and plants is dis-tinctly local to some small pocket of the entire protein-DNA machinery.

LEARNING ON THE OPERATING TABLE

My linking of consciousness with complex information processing is in ac-cord with this unconscious general-anesthesia scenario, in which mecha-nisms that would usually support such processing are no longer available: Brain waves have become slow and lumbering, and the latest, smartest parts of the brain have been turned off. So, does it follow that all sophisticated learning is unavailable when anesthesia strips us of consciousness?

Problematically, responses to anesthesia can differ markedly from one patient to another. On top of this, thankfully very rarely, some patients have remained awake during their operations and have been traumatized by their very conscious, very intense pain. So before testing for any learning, you first need to ensure that the anesthesia is sufficiently deep that the patient is truly unconscious.

The main method to establish that the patient is fully unconscious is by using EEG to ensure that the brain waves are sufficiently slow and deep, with little or no neural communication across regions. What can be learned after such checks have been confirmed? From the consensus of studies on learning in anesthesia, where testing only occurred after appropriately deep levels of anesthesia were carefully detected and monitored, there is no evi-dence at all of subjects, on waking, recalling anything from their operation. If word lists to be memorized are read out during the operation, patients have no memory of any of the words. If instructions were repeatedly given in the midst of the procedure, say, to lift a finger when the patient hears the name "Rumpelstiltskin," then nothing happens later in recovery when the experimenter gives the cue—patients do not lift a finger or recall having been given any instructions.

But, despite this, there are still some more subtle forms of learning that are possible. We actually learn all the time, in a weak sense of the word, with-out knowing that we've learned anything. For instance, as I mentioned in Chapter 1, if I say "artichoke artichoke artichoke artichoke artichoke" to you a hundred times in the next minute, you'll spend the rest of the day predisposed to react more to this word than usual. You'll read "artichoke" a little faster when you come across it, you'll notice it quicker in the supermarket, you'll think of artichoke a little earlier if I tell you to come up with as many vegetables as you can, and you might even be a little more inclined to buy an artichoke at the market. You won't be aware of any of this, and you'll have little control over it, but it happens nevertheless.

What occurs, from the neurons' point of view, when you've been exposed to such a word? The neurons that collectively represent "artichoke" in your brain were reactivated, and in the process, their thresholds for firing again were tweaked. They therefore will be a little quicker to draw their neuronal firing gun the next time you hear about or see artichokes. In one sense, this is akin to a micro-muscle getting a little exercise and becoming stronger from the use. But, actually, far more is going on. This little collection of neurons firing together a little more readily in response to "artichoke" is in fact an important predictive computation. The neurons are effectively saying: "Aha—artichoke is around again, perhaps it's now a bit more important and frequent, and we should reflect that fact by getting faster and louder to respond to it, or even anticipate it more keenly." Every time we are exposed to virtually anything whatsoever, this neuronal fine-tuning occurs.

There's no doubt that our unconscious minds are bubbling, spitting cauldrons of computations, based on this constant stream of neuronal tweaks that are dedicated to predicting the world around us. Our sensory perception is not a direct mirror to the world outside, but a series of computational steps designed to give our conscious minds the most pertinent information available. We move our bodies based partly on rich feedback from our senses as to where our limbs are and where the objects we're reaching for are placed in relation to our limbs. All this occurs seamlessly, unconsciously. Many of these low-level lessons are hardwired. We are primed to predict that the sun is above us, that objects have edges, and so on. This hardwiring is the product of many millions of years of evolution and designed with ruthless efficiency, so that we humans, and our mentally simpler ancestors, can swiftly extract the dangers and delights of the environment—and react before the predator or a competitor does. Many other unconscious lessons may once have been born as conscious explorations of the world when we were infants, but now they are so embedded into our world picture, and so buried under a mountain of more meaningful ideas, that we never consciously acknowledge their existence.

In one sense, this unconscious machinery for predictive learning is indeed complex, but only because of the sheer number of simple statistical calculations occurring, not because of the grand truths that are being unpeeled before our eyes. These neuronal tweaks that occur under the surface can only ever be the servants of our understanding, because only when their largest lessons are combined in consciousness can we really learn the interesting, deep patterns of life.

Just how limited is our unconscious mind, when applying these simple neuronal calculations? If you tell a patient recovering from an operation to complete the partial word "ash," she may be equally likely to say "ashcan" or "ashtray." But if you repeated the word "ashtray" when she was deeply unconscious, she will in fact be more likely to say "ashtray" in recovery. She will have absolutely no knowledge that she has just heard the word "ashtray" an hour or so ago; nevertheless, some part of her brain will remember. Some family of neurons has been tweaked to reflect this recognition, and she will respond accordingly. In other words, she has been "primed" to repeat a word to which she has already been exposed. Examples like this show that at some very superficial, unconscious level at least, words can be noticed.

But there are multiple features of a word—for instance, there's the sound, the grammatical features, the linguistic relationship a word has to others in various ways, and there's also the meaning of the word. Meaning is the first main level at which our minds structure information, since meaning requires relationships between items in a dense web, hierarchies of categories, and so on. In this way, meaning is an especially pertinent test for the argument that consciousness is equated not only with information processing, but especially with structured information.

Can patients be primed for the meaning of the word as well as the sound? For instance, if asked after the operation to complete the same partial word "ash," are patients more likely to say "ashtray" instead of "ashcan" if they were exposed during the operation to the word "cigarette"? If so, this would mean there had been a deeper triggering of activity, so that the neuronal population had spread from the sound of the word to activating any neurons coding for a related meaning. This is precisely what some researchers tested, and it turns out that, when under a sufficiently deep anesthesia, this form of learning is beyond us.

So, when profoundly unconscious, as happens under deep levels of general anesthesia, we can faintly learn under the radar of our consciousness that a certain word has just been presented to us, but that's about the limit of it. Anything remotely more complex, such as a word's meaning, is beyond our unconscious selves and requires at least some level of awareness. And, of course, for anything that we are actually conscious of learning—creating a strategy, memorizing a list, learning from instructions, or any of the myriad forms of information we manipulate every day—consciousness is certainly required.

Unconscious Better Than Conscious?

General anesthesia is the gold-standard method for studying just what our unconscious minds are capable of, because it puts us in a situation where we are fully unconscious, even though our brains are otherwise quite healthy and capable. Therefore, the limitations on unconscious processing described above should be taken as definitive. But science is always improved when it adopts multiple approaches to examine the same question. Indeed, various other techniques have been used to explore unconscious learning.

For instance, what about when we're already awake? Could it be that the deep ocean of our unconscious minds, when fed material from our conscious gaze, can churn over vast swaths of information and quickly allow conclusions to surface that are far more advanced than the insights that our deliberate, slow, conscious cogitations could ever produce? A Dutch researcher, Ap Dijksterhuis, carried out a series of experiments to argue just this point. He claimed that there were many situations where we should follow our gut instinct, where the slow, integrative processing of our unconscious minds was vastly superior to our clunky, far more limited conscious thoughts.

In one of Dijksterhuis's experiments (there are now quite a few, all of a similar mold), volunteers were asked to rank four imaginary cars in order of preference, based on a set of attributes they were given, one at a time. One car had 75 percent of its features set as desirable (for instance, "The Dasuka has cupholders"); another two cars were neutral, with 50 percent of facts good and the other half bad; and one car was the worst of the bunch, with only 25 percent of the attributes set as positive. Once participants were shown all the features for all the cars in turn, they were split into two groups. One group, the "conscious" group, spent 4 minutes thinking about all the attributes they'd just read about and tried to work out the most accurate ranking of the cars. A second group, the "unconscious" group, spent 4 minutes being distracted by anagram puzzles instead. This, according to Dijksterhuis, allowed their unconscious minds sufficient time to process and integrate the facts unfettered by their consciousness, which was adequately engaged in an irrelevant task. In fact, so Dijksterhuis's theory goes, the more the conscious mind is distracted by some other task, the better it is for the unconscious understanding of complex information. In the final, crucial stage, after the 4 minutes were up, both groups had to pick their favorite car.

According to Dijksterhuis's results, if there were only 4 facts per car, and 16 in total, then both groups did pretty well at choosing the best car, with the conscious group marginally on top. His interpretation was that, with few items, our conscious minds cope fine, because their very limited memory resources aren't too taxed by the moderate weight of facts to assimilate. But if you make the task vastly more difficult and have 48 attributes in total, 12 per car, then only about a quarter of the conscious group chooses the highest ranked car correctly—in other words, no better than a totally random guess. Meanwhile, a tremendously healthy 60 percent or so of the unconscious group makes the correct choice. This is a staggering difference. It led Dijksterhuis to conclude that in almost every sphere of decision making, be it political, managerial, or whatever else, "it should benefit the individual to think consciously about simple matters and to delegate thinking about more complex matters to the unconscious."*

There is one small problem with Dijksterhuis's conclusion, though: It is utterly wrong. If you stop and think about the task, as if you were a participant, it isn't difficult to see why. So you've just sat down in the small, stark testing room in front of a computer monitor. The environment is rather alien

---

*This result has now been loudly amplified within popular culture, with newspaper articles written all over the world, including titles such as "Want to Make a Complicated Decision? Just Stop Thinking." It also is the major theme of Malcolm Gladwell's book *Blink: The Power of Thinking Without Thinking*, in which he argued that immediate instinctive decisions can often be superior to long deliberation.

and you know that this is a psychology experiment, which makes your pulse race a little faster. What are these weird scientists going to test? Are they really going to look at what they will claim to be looking at, or will you be tricked in some way? Are you going to appear stupid? Socially awkward? A man with only half a brain? You're told to pick the most desirable car out of a bunch of facts, so you immediately try to do your best. First you're told, "The Hatsdun is very new." That's good, isn't it? Or does the car need to be broken in? Are some positives better than others? You haven't been given any instructions about this, which is disconcerting and confusing. You haven't seen any other attributes yet, so what are you meant to do? You decide the Hatsdun is provisionally in the lead anyway. Then you're told, "The Kaiwa has a poor sound system." Okay, that's bad, the Kaiwa is now at the bottom of your list (even though you only listen to the stereo for the news, and don't really care about the quality of the sound system—but you'll ignore this thought for the moment). It all feels a little overwhelming, with 48 facts to try to remember, but you do believe after a short while—only about 10 facts, actually—that one car clearly has accumulated more positive facts than the others. You can't see why you need to see all the facts to make your decision, but you keep paying attention and thinking about your choices, ready to modify them at any stage, and by the time all 48 facts have been presented, you've made up your mind. You could give your decision now, and start to open your mouth helpfully to volunteer the information—but that's not what the experimenter seems to want. They get you to delay your decision by doing anagrams, maybe to distract you, but you make sure you keep remembering your favorite car throughout, giving yourself a little reminder every so often. It's the same if you were asked to think for 4 minutes. It's actually pretty pointless, as it doesn't change your mind, as you'd already decided somewhere in the middle of seeing these facts. But you do what the nice researcher wants, so you get your money, or your course credits, and keep him happy.

This kind of experiment, as the above scenario illustrates, doesn't have much to do with the unconscious mind at all. It is a distinctly conscious task, and in a very similar study from another lab, this time carried out by Laurent Waroquier and colleagues, 70 percent of subjects reported explicitly applying various strategies while they were being presented with the facts to make up their minds, and had already made up their minds before they had seen all the facts. In other words, they admitted that their minds were fixed by the time the facts were all presented, with the other 4 minutes of distraction or deliberation being totally superfluous. So as an experiment designed to tap the unconscious domain, the test disastrously missed its target.

Furthermore, it failed the critical scientific test of replicability. If you add together all the studies that use this design, on average they find absolutely no advantage for unconscious processing. In fact, even some of Dijksterhuis's own experiments failed to find an unconscious-advantage effect. Other researchers, including a colleague of mine, Balazs Aczel, with whom I collaborated on an attempted replication, have found, if anything, that the opposite result was true: that even with long trains of facts, conscious deliberation provided an advantage over distraction. This is probably because being distracted, which fills up your conscious memory with other items, is more likely to dislodge your previously set correct memory for the order of the best cars. These dissenting papers, unfortunately, will never be splashed on the covers of any newspapers. But to me, the beauty and rigor of a sequence of carefully controlled experiments, doggedly seeking the truth, are far more exciting and fascinating than many sexy papers that spark an inferno of media interest by their novel, unreplicated results.*

So again, the message is clear: There's absolutely nothing superior about the forms of learning possible when only our unconscious minds can perform the calculations.

---

*Another reason, possibly, that Dijksterhuis's study has lent itself so easily to media coverage is that it appears to explain our "eureka" moments, when flashes of insight seem to appear suddenly, as if by magic, from the darkest depths of our unconscious. But is that really how it is? If our insights were truly unconscious, all we'd need to do would be to frame a complex problem, then stick our pens on some clean white paper and let our unconscious minds take control. Of course, if we did this, we'd be left with a dense nest of random lines, and we'd certainly be no nearer to the solution. Instead, most sparks of insight require a heavy investment of conscious thought. We need to develop strategies for exploring the terrain of possibilities, we need painstakingly to try each worthy permutation of the multitude of parameters, and finally, we need to have sufficient *conscious* understanding of the field to know when we've actually arrived at the solution. It's true that sometimes merely diverting ourselves, going for a walk, or having a nap seems somehow to dislodge our thoughts and make insights more likely. But is that because of the unbridled power of the unconscious, or instead because the break has allowed us to consciously take a fresh angle when we return to the problem?

FEELING YOUR WAY TO KNOWLEDGE

A more established and reliable body of evidence in this field involves performing some surface task, and unconsciously, accidentally, absorbing the underlying information within it. If I write the phrase, "Yesterday, the prime minister drive past me in the street," we all would know that it was grammatically wrong, but many times, with grammar, words just *feel* wrong, and we aren't necessarily clear about which explicit rule has been violated. It is possible to experimentally reproduce this impression that we can know that some item is correct or incorrect, without knowing why. The main format for such experiments is for researchers to show subjects incomprehensible sets of long letter strings—for instance, "XMXRTVM." The experimenter then tells these poor volunteers, rather sadistically, to memorize every one they see. After this part of the test, the participants are informed that the order of the letters in each string obeyed a complex set of rules, a kind of artificial grammar, but they will not be told what these rules are. For instance, one rule might be that the letter strings will always end with an M if there's an R in the mix. The subjects will now be shown more letter strings and have to say whether they obey the same rules or not. Even though subjects think they are entirely guessing much of the time, it turns out that they still get significantly more answers correct than they would by chance. It's nowhere near 100 percent, but then again, we had years to learn our native language's grammar with many thousands of examples, and these people just have a few dozen instances over 20 minutes or so.

Just how complex can these rules be? One of the most impressive examples is that we can unconsciously detect the connections between musical notes (when told just to memorize them in the same way as in the experiment above) and transfer that learning to the same connections in letters, all while still believing we are guessing randomly—but again I should stress that this learning is statistically above chance, but not by much, hovering around 60 percent (where chance between two choices is 50 percent), so it's miles away from infallible knowledge.

What does all of this show about the unconscious mind? It falls far short of showing that we can learn relatively complex information *in the absence of consciousness*, and for very similar reasons to the paradigm by Dijksterhuis shown in the previous section.

In order to learn anything at all about the strings of letters, we still have to be very conscious of the letters and attend closely to them when we first see them. If we're distracted, so that we are no longer reading the strings—even if we're staring right at them—then we learn absolutely nothing.

What's more, when we're asked just to memorize the letter strings in the first half of the experiment, most of us will be doing a lot more. Because of our exceptional consciousness, we are constantly looking for patterns; so in this instance, we just can't help noticing things like, "These strings always end with M—I wonder if that means anything." These thoughts quickly fade from awareness as we're trying to perform the main task of memorizing these difficult sets of letters, but perhaps these observations converge in our minds and allow us, consciously, to learn a little bit of the grammatical structure at a time, to pull us above chance. So, in other words, as with the Dijksterhuis experiments, with this experimental setup we cannot rule out the very real possibility that the structure is being learned consciously, at least to some degree, even if we later forget the conscious progress we originally made.

At the same time, these kinds of experiments are certainly not arguing for an unconscious advantage. There's no doubt that, had the initial instructions for these letter strings been to attempt to spot the rules that govern the sequence of letters, instead of simply to memorize them, subjects would have been considerably better at later deciding which letter strings obeyed the initial rules.

In many instances in life we at least semiconsciously build up these little observations, these tiny chunks of knowledge. Afterward, when we are not consciously recalling the rules, but only, at best, feeling a sense of familiarity, the initial groundwork nevertheless allows for proficiency on the task. Again, tennis is a good example. We start by consciously practicing the forehand stroke, section by section, and eventually, when we're proficient, we're no longer aware of all the little rules that make up what we're doing—all we need be aware of is the overall chunk: forehand. In fact, we might have virtually no memory of the learning process, but that has no bearing on our quality of play. Consciousness, therefore, may be required for any form of complex learning to occur, but once we've assimilated the task and the program is written, we merely have to execute it automatically. And if the heavy hardware of consciousness is accidentally reengaged on the habit, it often merely gets in the way.

And, of course, if the structure of some collection of information is particularly complicated, or especially hierarchical, then we absolutely need full-on consciousness in order to spot these patterns and learn them. Studies have shown that the unconscious mind is unable to cope, for instance, with most logical operations, with grasping cause and effect, with almost any type of sequence, with any mathematics beyond things learned by rote, such as multiplication tables, or with any of the various social and cultural facets of life we need to acquire in order to succeed.

So are there *any* sorts of tasks that the unconscious mind is unequivocally superior at performing, compared with the conscious mind? There are a collection of experiments that match this remit, albeit in a somewhat disappointing way: The unconscious mind can be superior when we consciously go a little overboard in our search for patterns. Just as pigeons can show superstitious behavior, by dancing for food that is delivered regularly regardless of their actions, so can humans entertain superstitious beliefs. But we can take this to exceptional levels, not only with astrology beliefs and the like, but also in more formal ways in the lab.

Imagine that you're sitting in front of a computer monitor, and you have to predict whether a light will flash on screen to the right or the left. That's the only task there is, but there are very many trials. In actual fact, eight out of ten times the lights will flash on the right, but in a random, unpredictable way. What's the best strategy here? If you've learned this pattern optimally, you'll respond to the right every single time, and end up with a score of 80 percent. That is what animals such as rats do when a correct answer leads to a food reward. But it is *not* what most humans do. Sometimes we as a species are just too clever for our own good. When we perform this task, we constantly look for some secret pattern, some hidden message in the apparently random stream. So, first of all, we lose marks by testing quite a few convoluted ideas about the pattern of the lights, most of which are wrong. Then, when we roughly understand that the lights flash on the right 80 percent of the time, we decide to second-guess it, and 8 out of 10 times we predict the right, but 20 percent of the time we predict the left. This one strategy actually lowers our score to 68 percent on average, because one-fifth of the time we're picking a side very unlikely to elicit the light flash. So the lowly rat outperforms humans on this task by a healthy 12 percent. What's more, if patients with damage to the lateral prefrontal cortex attempt this task, there is provisional evidence that they don't generate the usual wacky, wrong theories but instead perform optimally, just like the rats. This brain region is probably the area most intimately associated with awareness.

In many ways, though, this form of unconscious advantage is a cheat, because researchers could easily explain statistics and probabilities to the volunteers and encourage them to modify their behavior, so that they instead would behave optimally and their conscious abilities would catch up with their unconscious minds.

But to me these experiments highlight a key detail: These spurious suppositions are a litmus test for the extent of consciousness. Although these mistaken beliefs are essentially failures of our cognitive system, they are also a reflection of just how many patterns we *can* spot, and how successfully we *can* understand the world around us. So when the pigeon twirls around for food, that is potentially a marker of minimal consciousness, and when a human believes in astrology, that is actually, strangely, a solid indication of a substantive form of consciousness—no other type of animal could have such a spectacularly irrational schema as astrology, and there is also no chance for our simplistic unconscious minds to generate these intricate, rich false theories. Such elaborate failures as astrology, or, more pathologically, conditions like schizophrenia, are perhaps the price we pay for our exceptional consciousness, our deep hunger for wisdom, and our incredibly successful ability to detect profound *real* patterns, greatly furthering science and technology in the process.

EAT POPCORN: THE UNCONSCIOUS TAKES CONTROL

We've arrived at the conclusion that the unconscious mind is the dullard at the party: There may be a few genuine experiments that appear to show an unconscious advantage, but these cannot rule out the initial powerful gaze of consciousness drawing out the meaning in the string of data. If this isn't the explanation, then instead any apparent unconscious advantage can be explained by consciousness occasionally overactively latching onto a specious pattern—a potentially temporary glitch that our uncreative unconscious isn't capable of generating.

And as the evidence mounts that unconscious processing is limited to simpler thoughts, with little or no structure, the argument for the role of consciousness in complex information processing, especially involving layered meaning and deep patterns, is in turn strengthened. If we are to absorb anything complex, we need to direct our attentional gaze firmly in the right direction and consciously identify the relevant features.

But the question of what we are consciously or unconsciously capable of learning is quite distinct from which of these sides of our minds usually controls our choices. So who holds sway over the lion's share of our decisions in life—our conscious or unconscious minds?

The most famous early experiments attempting to answer this question investigated whether our decisions can easily be manipulated by subliminal messages. This field was spectacularly brought to the public eye in 1957 when James Vicary, a market researcher, conducted a simple experiment, which he related in great detail. While cinema-goers watched a movie, brief frames were inserted for a tiny fraction of a second (too fast to spot) ordering the viewers to "EAT POPCORN" or "DRINK COKE." Vicary related how these instructions appeared every 5 seconds throughout one film, every time it was presented over a six-week period, to 45,699 participants in total, at a movie theater in Fort Lee, New Jersey. Vicary claimed that these messages raised sales of popcorn by 57.7 percent and Coke by 18.1 percent. A year later, the CIA produced a report describing how to exploit this important result for their own aims, and soon such subliminal additions to media were deemed illegal in the United States. Meanwhile, the notion that we can be heavily influenced by such brief flashing instructions was firmly fixed in popular culture. Unfortunately, Vicary never properly published these results, and five years later he freely admitted that he had totally fabricated the data! In fact, despite considerable efforts to replicate these original bogus findings, there's no evidence that such subliminal messages influence our behavior.*

---

*These actions by the CIA and the U.S. government are themselves iconic examples of poor, knee-jerk decision making, and they emphasize how choice quality can be improved by pausing, deliberating carefully and consciously, and relying as much as possible on proper scientific data, such as peer-reviewed publications, with good evidence of effects being repeated in different labs.

But unwittingly, Vicary was, in some ways, very much on the right track. We hardly need science to tell us that our decisions rarely resemble ideal, enlightened, conscious choices. If we could make our own decisions in an entirely objective way, as if we were deliberating slowly and carefully on behalf of a complete stranger, we might always be able to choose the long-term constructive option and never opt for the short-term destructive one. We'd never overeat, always exercise properly, never be ruled by anger, always pause and analyze the alternatives, and avoid impulsively choosing the toxic route. We would weigh decisions based on the best scientific evidence, instead of a hunch—or, worse, a biased, dogmatic fragment of conjecture—and we'd always inject an appropriate level of skepticism into our decisions, instead of rushing headlong into believing a completely irrational suggestion. Unfortunately, even the best and brightest of us chronically fail to live up to this ideal. Why are we so very bad at making decisions?

In the main, our paths in life, from the trivial to the momentous, are heavily shaped by evolution. We can't escape our biological heritage, which has been designed, over many millions of years, to keep us alive. We overeat because normally, in nature, food is scarce, so when there is a plentiful supply, the desire to "stock up" is incredibly powerful. We suffer heavy stress, even when there is not even a remote threat to our lives, because we are built to strive desperately in a dangerous world. We are also engineered to impress—to rise socially as far as we can—partly to secure more resources, but also to find sexual mates. Sex is one of the main driving forces of adult life, because, after all, passing on our own brand of genetic ideas is the main evolutionary "purpose" of our existence. (It is also the area where we probably make the most profound, life-shaping mistakes, from choosing the wrong person to marry, to having affairs, and so on.) In all these decisions, it regularly feels as if we have little say in the matter—that we simply feel these urges and act on them, like a passive traveler within our own bodies. We may even aggressively question our poor decisions, but feel powerless to stop them, or else supplant these doubts with fatuous, after-the-fact justifications, which nevertheless somehow seem to pacify our pricked conscience.

Science puts some meat onto the bones of this impression that our unconscious minds are the engine room of decision, with some stark, disturbing results.

When I was an undergraduate, I learned of many fascinating experiments, but there was one seemingly trivial, straightforward study that I actually found most striking, even disturbing, because of its widespread implications for human psychology and free will. You are told that you will have to perform an exercise—sit against the wall without a chair, with your thighs and calves at right angles. This is easy to begin with, but after a minute or two, even the muscles of very strong, fit people will start to burn. You're informed that you're actually earning money for a relative, and the longer you hold your position, the more money they will get. The cash sums are pretty trivial, equating to perhaps 30 cents every 20 seconds, but that doesn't stop volunteers from trying hard to earn a bit of extra money for their families. Now we share half our genetic identity with our offspring, parents, and siblings; a quarter with our grandparents, aunts and uncles, and nephews and nieces; and an eighth with our first cousins. The people in these studies, carried out across cultures and including both men and women, all behaved the same way—they held their painful position for an amount of time that related to the closeness of their relative. So the amount of time they could hold the position for themselves was the longest, followed by the time it was held for siblings, then nephews, etc., and finally, the shortest time and most measly payout went to cousins. Volunteers were not at all aware that they were following this pattern, but some part of their unconscious minds was presumably making an assessment of how many genes their relative shared with them, and based on this knowledge, calculating how much suffering to apportion for each relative's reward.

Countless examples, in a similar vein, exist to demonstrate the frightening extent by which our choices can be unconsciously, irrationally influenced. The Israeli Nobel laureate in economics and psychologist Daniel Kahneman knows better than most just how many complex, potentially competing sources can, under the surface, contribute to any eventual decision. When he was a Jewish boy of around seven years old in Nazi-occupied France, he once was distracted playing with a Christian friend and accidentally stayed out past the 6 p.m. curfew for Jews. Turning his sweater inside out to hide his Star of David, he began to walk home quickly. Then he saw a soldier approach—and not just any soldier. This soldier was wearing the ominous black uniform of the SS, the notorious Nazi paramilitary force that carried out many of the atrocities of the Holocaust. And he was staring intently at the young Daniel, who was speeding up to try to walk past him—but in vain. The SS soldier beckoned the boy over, but instead of subjecting him to the fate of other Jews who had broken the curfew, he picked the boy up affectionately, hugged him, and spoke warmly to him in German, completely failing to notice the thick Jewish symbol on the inside of his sweater. He showed the boy a picture in his wallet of his own son, and even, in an ultimate act of irony, gave this Jewish boy some money to help him on his way. This event firmly placed Kahneman on the path of psychology as he sought to understand the many sources of our choices.

A few decades later, by now a well-established psychologist, Kahneman coauthored, with his longtime collaborator, Amos Tversky, a classic, damning, and comprehensive paper outlining a variety of ways in which our everyday thoughts and decisions can be biased, revealing them to be facetious and arbitrary on a regular basis. For instance, say volunteers are asked the question: "What is the percentage of African countries in the United Nations?" They are then shown a kind of wheel of fortune, which is swung to land on any number between 1 and 100. This wheel has nothing to do with anything, really—it just spews out a random number that the volunteers observe. But if the wheel lands on the number 10, then the volunteers' median estimate becomes 25 percent of African countries, and if it lands on 65, the median estimate is 45 percent. This unconscious anchoring of our choice of answer is entirely random, but nevertheless very significantly affects our guess.

All of this indicates that consciousness might well be designed to make those big innovations, and uncover those deep patterns to the world, but that it is, by default, more a slave to our decisions than its master. Biologically, we are built to survive and reproduce, and the unconscious part of our brains does all it can to steer us toward these simple goals, with our conscious minds usually just a superlative servant of those unconscious aims.

UNCONSCIOUS DECISIONS AND THE FREEDOM TO CHOOSE

Although there is a bleakness to this perspective, there is some research to point to an even more disturbing picture, one where all decisions are necessarily unconscious, even when we believe them to be made entirely within the remit of awareness.

In the early 1980s Benjamin Libet demonstrated this fact with sufficient clarity that it has unnerved psychologists and philosophers ever since. Participants, while being scanned with EEG electrodes, were asked to lift a finger spontaneously, whenever they wished, and state exactly when they had the urge to do this. Because talking is another kind of movement, they couldn't just say "now" to indicate the timing of their decision, as this would have its own movement initiation process, meaning all answers would be delayed by the time it took to say "now"—probably an extra third of a second. So Libet designed a cunning means by which subjects could accurately report the time of their decision without affecting the rest of the experiment: Subjects would watch a special clock that had the usual circle of numbers, but only a single rapidly rotating hand. When they decided to act, they would note the nearest number to the clock hand at that precise time. Then, after completing the movement, they would report this number.

Using this special clock method, Libet could now calculate when each conscious decision to move a finger occurred. He first found that the volunteers were making this decision about a fifth of a second before they actually moved a finger. No surprises there. But the EEG electrodes picked up a strikingly different story—brain activity relating to the preparation to move a finger began to ramp up a good third of a second, on average, before the subjects consciously believed they had decided to move—and sometimes this brain marker was detected a whole second beforehand.

Does this mean that even our simplest, most humdrum choices are decided in our unconscious minds, and we are mere spectators, only believing that we are in control because our brains are designed to make us think like this? Many scientists and philosophers have tried to wriggle out of this uncomfortable conclusion. The main author of the above experiment, Libet, claimed that we still have free will, because we always have the conscious ability to veto any internal choices we make. Unfortunately, there is an increasing weight of evidence tipping the scales toward a similar unconscious source of even this decision to veto our responses.

Another lunge to save free will came from the philosopher Daniel Dennett, who claimed that the problem went away if you assumed that we were measuring things too accurately. If I were to try to explain where Cambridge University was located, for instance, it would be ludicrous of me to pinpoint it as the stone crown above the front gate of Kings College, even if this is a relatively central location of one of the grandest university buildings. Instead, the university is located via the hundred or so college, department, and administrative buildings strewn around Cambridge city.

Similarly, Dennett argued that all the Libet experiment really showed was that consciousness was smeared across time, perhaps of the order of half a second long, and that it's entirely invalid to assume it has a single precise temporal location. This is in some ways a very plausible idea—consciousness involves a massive collaboration among a multitude of large brain regions and is undoubtedly a complex, perhaps even slightly lumbering, process. So the suggestion that consciousness has a somewhat nebulous timescale would make perfect sense.

A recent computational model by Stanislav Nikolov and colleagues for how the brain recognizes its own important neural events, over and above the mere random chatter of neurons, provides a detailed justification for Dennett's position. Nikolov's model showed that it was actually counterproductive to detect a decision when brain activity for the decision was just on the cusp of rising above the usual random baseline hum. Neural activity is always rising and falling, because the brain is a noisy, semichaotic place. What the brain needs to do is to carry out some solid, cautious statistical tests on its own activity, and only trust that a decision has been made when collective activity is quite high, clearly above chance—and this point is reached considerably later than the point at which this quiet ramping-up process begins. Otherwise, the brain would constantly be misinterpreting chaotic noise as significant, meaningful activity.

Imagine, as an analogy, that you are standing in the crow's nest of an early nineteenth-century military ship, in the midst of war. It is your task to alert the crew that an enemy vessel is approaching. One strategy might be to get excited by any movement on the horizon, and shout to the captain far below that there may be an enemy approaching. But what if most of the time after relaying these terrified alerts, you actually found that you'd spotted a bird, or a jumping dolphin, or a particularly high wave, or that there was just a speck in your eye? The whole crew would be pushed to battle stations a hundred times a day, and would be too exhausted to fight when that dangerous frigate did in the end turn up. Instead, holding fast to the lip of your crow's nest, you set a criterion: You say to yourself that you will only alert the captain when you are reasonably sure that an enemy ship is detected. This rule might cause you to lose a few seconds of preparation, but the battle stations will most likely only need to be set that one time, in a few days when they are actually needed, saving much collective energy.

Likewise, returning to the Libet experiment, where you apparently decide to lift your finger considerably after your unconscious brain chooses to, Nikolov's computational model argues instead that the brain simply cannot tell that a true decision has occurred until activity has passed a high enough threshold, possibly at exactly the time when the subjects say that they decided to lift a finger. In other words, there may be no unconscious decision to speak of—it's just that the decision only qualifies *as a decision* when neural activity is sufficiently high.

In a far more recent imaging experiment, now using fMRI, John-Dylan Haynes and colleagues adapted Libet's experiment to show that brain activity in the front section of the prefrontal cortex could reliably pick out the difference between whether we've chosen to initiate a left or right finger movement. This in itself is exciting enough, but what's even more remarkable is that this activity was detected up to 10 seconds prior to the conscious decision.

Although, conceivably, the same defense could be applied here—that for most of these 10 seconds the increase is too indistinguishable from random fluctuations in the brain to reliably call it a decision—that now seems quite a stretch. Clearly, more research needs to be carried out in order to untangle this issue. But the possibility that every single one of our apparent conscious decisions was actually previously fixed in our unconscious minds needs to be given some consideration.

Provisionally, though, there are multiple potential routes for consciousness to play a key role in our decisions, particularly if we look not just at simple, arbitrary choices, but the whole gamut of our decisions. One important distinction again is between habitual choices and novel ones. For some routine, repetitive decisions, which seem quite automatic and potentially unconscious, there may well have been some initial careful conscious thought to set it up that is now long forgotten—analogous to the distant memory of learning the specifics of forehand in tennis. For instance, I give little thought to what I eat for breakfast out of a few choices, but that's partly because I at one point in the past *did* give this some conscious assessment and informal experimentation, so that I landed on a set of healthy choices that I enjoy eating. Personally, if there is some routine that initially started as an arbitrary gut instinct, then I would be worried about that habit and wish to revisit my choices.

But what of novel decisions? It's almost a waste of resources to apply consciousness to decide when to move your finger in an experiment, for instance. In contrast, there may be many hours of clear conscious thought put into choosing what degree to study, or what career to move into. If these were written in a diary, then you could even clearly see the evidence for consciousness at work: the logical arguments, the weighing of pros and cons, and so on. Given that the unconscious mind cannot operate on these structured, highly refined levels, there is a strong case to be made for consciousness being heavily involved in those important complex choices.

ASPIRING TOWARD FREE WILL

So although the Libet experiment has been classified as the quintessential psychology experiment to refute free will, its implications may not extend much beyond the simplest of mental decisions, and its conclusions may be further weakened by taking into account the cautious statistical approach required to detect neural events in a sea of random noise.

But leaving this debate aside, I see the issue of free will rather simply. If the question is whether we have the freedom to choose, independent of the machinery of our brains as it interacts with the world, then of course we don't have free will. *We are that machine*, so how could we be independent of it? If one believes that no machine could ever have free will, then we don't either, and that's the end of the debate.

But we underestimate the detail of our neural machinery at our peril. We are unimaginably complex, with around 600 trillion connections inside each of our brains. And this particular machine is very special because it is an immensely powerful information-processing device, with a tremendously rich internal model of the world, along with accurate copies of various events from its past. These fascinating properties give the enticing, persuasive illusion that we can make decisions independent of the outside world, or even outside of the fettered constraints of our neural hardware. For instance, imagine that, by some quirk of fate, I lost all my senses, and medicine kept me alive by various feeding and breathing tubes. I would lose all external input, but I might still have an active conscious life for many years—I might mentally write the odd novel, compose some music, or generate some naive theories about politics and philosophy. What I would do in my own internal world is hard to predict exactly, but there could be any one of a million paths as I creatively explored the knowledge I'd accumulated up to the point where my interaction with the outside world was cut off. No artificial machine could come remotely close to the range of possible, unpredictable activities that might ensue if a human lost input from the outside world.

I'm not arguing here that the sheer complexity of our minds somehow allows free will after all. Instead, I merely want to highlight one reason why the illusion seems so compelling. As it is, there is little we can do to shake the embedded illusion of free will, so we may as well play along. Part of the game we play is to redefine free will more softly in terms of decisions that are consciously, rationally made under normal circumstances. For instance, a person acting on the delusions of his schizophrenic illness would be assumed to have a substantially diminished free will, because his insanity robs him of the chance to make those conscious, rational decisions that we take for granted.

Even on this level, though, there are strong grounds for giving many people, under many circumstances, the benefit of the doubt, when we might initially have assumed they committed some wrong. A substantial proportion of our everyday decisions have an unconscious bias aggressively pushing us toward a selfish, short-termist agenda. And, as neuroscience progresses, we are increasingly realizing that thought patterns and behaviors that previously would have been classified as "personality problems" are actually forms of mental illness, with detectable genetic and neurophysiological roots.*

Unfortunately, as a species, we do have a vast potential for destructive acts. This toxic capacity is partly due to the supercharged conscious component of our brains, which can be engaged ruthlessly and innovatively to achieve our irrational unconscious goals. But at the same time, the enormous

---

*And if we do subscribe to the idea that no one has any free will in the stronger sense, because we are all just machines, then one logical consequence of this is to forgive, or even simply and fully to accept all the actions that everyone makes, just as we might accept natural events from inanimate sources. This might not be a realistically attainable perspective, but it would certainly be one free from anger, intolerance, and hatred.

analyzing capacity of our consciousness also provides the potential to overcome these default limitations. Our conscious minds are uniquely capable of analyzing the consequences of our choices, even though, by default, many decisions may be made with little conscious input. The trick, somehow, is to wrestle control, as much as possible, into the realms of consciousness.

One strategy may simply be to use our conscious minds more often: Our lives could improve considerably if we tried to ensure that any remotely important choice was given the full force of conscious, rational deliberation. By trusting the sophisticated conscious space we have, we can bring to bear the most refined, capable computational aspect of our brains, understand our own goals, avoid inherent biases, and make far more enlightened decisions.

One aid to this endeavor would be to elucidate the psychological and neural landscape of our conscious minds so that we could better understand the scope and limitations of awareness. Although this chapter has begun to make some headway toward this goal by exploring the distinction between conscious and unconscious processes, in the following two chapters I will more directly elucidate the psychology and neurophysiology of awareness.

4

Pay Attention to That Pattern!

Conscious Contents

Dangerous Daydreams

Occasionally, when composing a new piece of writing, I like to go for a stroll, as I find it useful for generating free-wheeling ideas. And so, to help cement a few concepts together for this current chapter, I set off on a leisurely mid-summer walk to Byron's Pool, halfway between my house in South Cambridge and the village of Grantchester. Named after the poet Lord Byron, who was said to frequently bathe at this picturesque little spot when studying at the nearby university, Byron's Pool has been a favorite, secluded swimming location for many generations of students, including Virginia Woolf and Ludwig Wittgenstein.

On the twisting country road, the signpost for the pool is quite prominent, and it's not as if I've never visited the site before. I know I should hit the turning about 15 minutes into my walk. But after 30 minutes, I somehow found myself at the Green Man pub in the middle of Grantchester, having completely missed the sign many minutes back. In fact, I could hardly remember the visual part of the walk at all—not Canteloupe Farm, whose name always makes me want to eat a melon; not the bridge over a round, widening bubble of the river Cam, surrounded by trees and statues of animals; nor Grantchester Orchard. Though I had been distracted once, very briefly, by a loud car horn on the road beside me, a few seconds later I was back in my own world, concentrating on the psychology of consciousness.

4
109

What occupied me most of all during this walk was how to piece together the structure for this current chapter, a few linked items at a time.

No matter, I thought, I would be bound to catch the sign on my way back, and at least visit Byron's Pool on my second attempt, about 15 minutes later. I shouldn't have been so confident. Thirty minutes after I set off on my return route, I found myself back in the city of Cambridge, near my house. I had succeeded in entirely missing the sign for Byron's Pool not once, but twice. The chapter structure was shaping up well, though, so I wasn't too disgruntled—just feeling a tad sheepish.

This wasn't the first time that I'd been completely oblivious to my surroundings while faithfully attending to the inner cogs of my mind. In fact, I seem to have a disturbing knack for entering my own world and utterly ignoring whatever my senses are brightly picking up right in front of me. When I'm driving on the motorway, or even in the busy city, it doesn't take much for me to be engrossed in replaying a memory or following a train of thought, while my perception of the road, of moving the steering wheel, the pedals, and so on, all disappear from consciousness. All I'm aware of at these times is the memory of that strange discussion the other day, or the words of a future paragraph I want to write, or the format of a new experiment I want to run. Somehow I do slow down and speed up in response to cars around me, and navigate through traffic lights—but all without the watchful presence of my conscious mind. In fact, I might also have the news on the car radio, or be "listening" to an audiobook, but these may as well be random noises that bash in vain against my ears, failing to penetrate very much further in.

In my defense, it never actually seems dangerous. As soon as anything unexpected happens, such as a car suddenly braking in front of me, I'm back immediately—aware of the car and my surroundings. I swiftly calculate what to do and act on those thoughts to negotiate the problem safely.

I know I'm not alone in my absentmindedness. In fact, there seem to be some people considerably more adept than me at crawling inside their own intellectual shells. The worst case I heard, from the mathematician Ian Stewart, was of the pioneer of cybernetics, Norbert Wiener, who, as well as being a very talented mathematician, was notorious at being distracted by his work and forgetting important details of his own life. When they moved to a new home, his wife, knowing exactly what he was like, wrote the address very carefully on a piece of paper for him and pleaded with him not to mislay the item. "Don't be silly, I'm not going to forget anything as important as *that*," he replied, safely putting the paper in his pocket. But later on, Wiener started obsessing over his latest insight, grabbed the nearest, most convenient piece of paper he could find—which happened to be his new address—and furiously scribbled equations all over it. Having found that the new idea had some serious flaw, he threw the piece of paper away in disgust.

When it was time at the end of the day to return to his family, he vaguely remembered something about a new home, but of course had no paper address anymore, and couldn't find it anywhere. His only recourse was to head back to his old house, and sheepishly see if any neighbors knew where he'd moved to. As he arrived, he noticed a young girl sitting beside the old house and approached her.

"Pardon me, my dear, but do you happen to know where the Wieners have mov—?"

"That's okay, Daddy. Mummy sent me to fetch you."

Attention Funneling Raw Data to Build Experiences

So far in this book we've arrived at the position where consciousness is a physical, brain-based process, most effectively investigated by science. While information processing and the management of ideas is at the heart of evolution, the extra, far more capable forms of information processing inside a brain allow consciousness to emerge from this exquisitely designed biological computer. But awareness doesn't arise in all species, or even at every moment in human life. In those species with the capacity for consciousness, low-level processing or routine actions are carried out unconsciously. Only when our data processing is of a sufficient magnitude and complexity, and of a certain type, does consciousness occur.

This and the next chapter will continue exploring exactly what forms of complexity, and what type of information processing, relate to consciousness, and how attention funnels the raw data we soak up from the world and converts a small portion of our input into the experiences that fill our lives with meaning. In this chapter I'll be centering on the psychology of consciousness, and in the next on how the brain creates our experiences.

As my stories about absentmindedness illustrate, attention is closely related to awareness: What I attend to is what I'm conscious of, and whatever falls outside of my attention is processed, if at all, by my unconscious mind alone.

But before describing the intricate psychological details of what attention is, and how it relates to awareness, I will pause and ask, from first principles, what the purpose of attention might be.

Attention addresses a basic data-processing issue that almost all types of computers face. Simple information-processing systems, such as plants or bacteria, are receiving only a faint trickle of information from their senses and have only a rudimentary ability to process that information. On the whole, all the information they receive, they process as far as they can. But for more complex systems, where the information stream flowing in is overwhelming the system's ability to process every item fully, there needs to be some decision process about which subset of all this mountain of data is most deserving of further analysis, and which other subsets are best ignored. This data-filtering and -boosting mechanism is attention.

At the extreme end of this continuum of information input and analysis capabilities sits the human brain, with enormous cortical regions capable of very deep analysis. Hence the necessity for highly aggressive attentional filtering and boosting. This filtering can be so intense and focused, for instance, that we can foolishly walk for 30 minutes and hardly notice our surroundings at all.

Human eyes have around 100 million photoreceptors, each of which can pick up about ten visual events every second, so our eyes are effectively receiving a billion pieces of information each second. If you include the information pouring in from our other senses, that's a staggering quantity of data for our brains to sift through every moment of our waking lives. This weight of input is hardly unique in the animal kingdom—after all, many mammals have senses at least as acute as ours. But humans come up trumps from the analysis point of view, as, relative to the rest of our brains, we have considerably more general-purpose neural real estate by which to carry out detailed processing on the data we receive. This creates a dizzying potential for learning. And, as our everyday lives illustrate—and the fruits of science and technology reinforce in a muscular way—the deepest, most forensic analyses tend be the ones that offer the most rewards.

If we had an infinite source of energy by which to crunch the numbers, and an infinitely fast brain by which to make the calculations, then there would be no problem, as we could analyze every scrap of data to its fullest capacity and never miss an opportunity or be caught by a threat. But of course, in reality, it takes time to process anything, and human brains consume a frighteningly large proportion of our body's total energy resources.* So we have to be ruthless in what we filter out, and tremendously picky about what data we allow through to the highest levels of processing.

One key question here is just how you know what is biologically relevant. Imagine that there's a poisonous snake near a plentiful food supply. An animal walks toward the food, and then quickly, effectively, its attentional system focuses in on the snake and little else. It recognizes the danger and flees. This has potentially saved the animal's life. But what if the snake were dead, and the animal were starving, with little other food around? If greater analysis could have revealed the snake to be no longer a threat, perhaps by smell, or, in the case of particularly smart animals, a tentative poke with a finger or stick, then the animal's life might have been saved by this extra understanding, which prevents it from running away and enables it to eat the food in this location.

This example shows that, ideally, you need an attentional system that is guided closely by various emotional and instinctive signals so that it can automatically—and quickly—focus in on any immediate threat. If threats aren't present, then attention should choose to center on other vital biological needs: food, procreation, social status, and so on. *But* for all these drives, there may be a better, slightly more lateral step to achieve them, if only we understood the situation a little more deeply.

In a broad sense, because of these potential alternate routes to optimization, almost everything could be biologically relevant, potentially, and if you

---

*Even though the human brain is a mere 2 percent of total body weight, in newborns this single organ requires a staggering 87 percent of the body's total energy. A five-year-old has a brain that greedily guzzles nearly half of all the energy the child consumes, and even in adults this figure is at least a quarter, though that proportion can rise dramatically if we've had a mentally taxing day—for instance, when studying for exams. In fact, some biologists have suggested that the energy demands and complexity of a human brain are nearing the endpoint of what is biologically possible and that if you started trying to cram even more neuronal wires into the brain, the additional miniaturization that this would entail would turn all brain signal into random noise—and the cleverest organ in the known universe would suddenly become one of the dumbest.

have enough brain resources, why not explore the informational terrain extensively? Humans thus readily attend not only to local external events, such as dead snakes, but to more abstract ones, too, such as the movement of the stars, or the factors that help crops grow, as well as to the internal information stream—for instance, an inner monologue about book compositions. Although such subjects have little to do with direct survival, on the surface, we continue attending *just in case* a deeper analysis reveals that they may assist us, if only we could reveal some hidden spark of wisdom from them. After all, our universally curious natures occasionally do strike gold, as exemplified by science and all of its technological products, which have made it so much easier for us to meet our basic survival needs.

Not Spotting the Wood, the Trees, the Birds, the Soil, the Flowers, the . . .

In my embarrassingly blind walk described at the start of this chapter, it felt to me that attention was the gateway to my awareness—without attention directed toward some feature of the world, I simply would not be aware of it. This intuition has been repeatedly demonstrated in psychological experiments.

The first, commonly referred to as "change blindness," is perhaps one of the most striking and important experiments in either attention or consciousness, and reveals almost all key features of both (see Figure 5 for an example). Its standard form, first carried out by Ronald Rensink and colleagues, involves two photos identical except for one feature. These are presented to the volunteer on a computer monitor, with a blank screen sandwiched in between for about 80 milliseconds. When I tried this experiment some years back, in a large seminar at an academic conference, both pictures were of a military plane at an airport. With lots of flicking back and forth between pictures, I was initially convinced that the experimenter had made a mistake and that the two pictures were in fact identical. It didn't help that the speaker, Bob Desimone, a world leader in the neuroscience of attention, cheekily suggested that the faster you spotted the change, the higher your IQ (to my relief, this isn't true!). It took me and most of the audience a long time—perhaps 30 seconds—before we noticed that one of the pictures showed the plane missing an entire engine. My attention, heavily influenced by my expectations of what was important in the picture, was moving from interesting feature to interesting feature, such as the national insignia that identified the plane, the soldiers in the foreground, at the plane exit, or coming down the stairway, and so on. It simply didn't occur to me to direct my attention to the engines, as you don't expect planes on runways to lack these.* And because I couldn't attend to more than around a few items for each of the two pictures, I never had a chance to notice the change. This staggering delay to spot obvious, blatant differences between viewings is common, and another example of how limited the scope of our attention—and awareness—can be.

A real-world version of this experiment, by Daniel Simons and Daniel Levin, shows even more dramatic results. Unwitting volunteers walking on paths in the university campus are asked by one experimenter for directions to a certain building. A map is produced, which the volunteer helpfully starts examining to work out the best route to the building. Rather rudely, a door is then briefly passed in between these two people by two other secret experimenters, and the person asking for directions is swapped with someone else. This new person is holding an identical map and continues asking for directions as if nothing has changed. Incredibly, despite the fact that the new person has a different face, voice, clothes, and so on, less than half the volunteers notice that the person has changed.† They are presumably so busy attending instead to their inner spatial worlds, as they work out the best route to advise the passerby, that they fail to attend properly to the man—or rather men—holding the map. And without attending adequately, they aren't aware of the change. In some ways this is the experimental crystallization of absentmindedness—a trait I far too readily demonstrate, for instance, when I attempt in vain to go on walks to Byron's Pool, or Weiner exhibited by being so distracted as to fail to recognize his own daughter.

It isn't merely changes that are missed if our attention is elsewhere, but also obvious anomalies in the world. In another famous attention experiment, known as inattentional blindness, originally carried out again by Daniel Simons and this time Christopher Chabris, volunteers watch a video

---

*To try examples out yourself from the original study, visit www.psych.ubc.ca/~rensink/flicker/download/.

†For a video example of this from the original authors, visit http://youtu.be/FWSxSQsspiQ.

of people playing with a basketball, the task being to keep a careful count of the number of passes. At some point in the video, someone slowly walks through the basketball players wearing a full-body gorilla outfit. Amazingly, only 44 percent of people actually notice the gorilla, despite the fact that it was in the video for a good 5 seconds, even pausing in the middle of the shot to face the camera and comically beat its chest.* Again, because people are deeply attending to something else—in this case the basketball passes, as well as their internal tally of these passes—they don't attend to, and aren't in any way aware of, the mock gorilla invading the scene. Such black holes in our awareness are bread and butter for pickpockets and magicians, who are experts at manipulating our minds so that we fail to attend to the loci of their tricks.

In all these instances, the story is the same: What we attend to equates with what we are aware of, and the boundaries of consciousness are extremely tight compared to the broad scope of stimuli entering our senses. This small subset of the world that we are actually aware of allows us to miss striking changes in a scene, or highly unexpected events occurring right in front of us. But the flipside of this phenomenon is that it helps us understand a few features of the world far more deeply than we would otherwise, and potentially to perform complex tasks in relation to them.

A BRIGHTER, MORE VIBRANT WORLD

So there is clear experimental evidence for the suggestion that we engage in tremendously aggressive attentional filtering in order to focus our awareness on a small component of our informational world. But what about empirical support for the second feature of attention, that it also acts to boost information processing?

If I'm trying to spot my wife in a crowded train station, I know she's wearing a red sweater and that she has black hair, so I concentrate on scanning all the people for red sweaters and black hair, and then, if I find a partial match, I look for her facial features. When I do this, it feels as if I'm barely

---

*For an example of the video used in this experiment, visit http://youtu.be/vJG698U2Mvo.

aware of the sounds around me, but every red item of clothing or head full of black hair stands out vibrantly. My whole mind seems attuned to redness, as if other colors are only important at that moment in a negative sense, because they are not red.

There is good experimental evidence to back up my impression that this controlled, directed attention is capable of actually enhancing both information processing and one's awareness of certain features of the world. This is trivially true when attention causes me to move my eyes toward some object of interest, thus allowing the central part of my visual field, the fovea, to focus on the item. The fovea has a particularly dense collection of photoreceptors and considerably more acute resolution than my peripheral vision. So the simple act of moving my eyes and focusing on an object allows a far higher, richer stream of visual information to be absorbed than if I wasn't looking directly at the object. Of course, once an animal has oriented toward an object, it tends to stare at it, soaking up the greater stream of input so that it can analyze more carefully what the object is, and thus further boost its processing of the information it is collecting for that object.

But you don't have to be looking anywhere near the object for the boost to occur. For instance, consider the following experiment. Say you are required always to stare at a dot in the center of a computer screen. On half the trials, an additional dot will then randomly appear very briefly near one of the four corners, while on the other trials there will be no such peripheral dot. When this dot does appear, you really pay attention to it in your peripheral vision, because you know from the task rules that this corner dot is warning you that a difficult-to-detect, very faint object is about to turn up in that location—and the whole purpose of the experiment is to notice these faint objects. So even though you are constantly staring at the central dot on the screen, as instructed by the experimenter, you somehow shift your attention to this quarter of space, to prepare for the difficult-to-see object that's about to appear. If you're ready and attending in advance to this quadrant, then, for this quadrant and this alone, your vision actually improves—you are conscious of fainter targets than normal, objects you wouldn't normally have seen, and you can also detect targets faster.

Not only this, but there's evidence that we actually see things more vibrantly, too, under the amplifying power of attention. If you attend to the location where a clearly visible object is about to appear, that object will be perceived to have more contrast than if you are not attentionally preparing for it to turn up at that location.

Attention, then, clearly boosts awareness. It allows us to consciously see things that otherwise would have been too faint to detect, and it really does make things appear more vivid.

THE ATOMS OF THOUGHT

But how does attention filter and boost the incoming signal and shunt this output into awareness? Although this chapter is primarily a psychological one, insights about the mind can often gain clarity if examined in the context of corresponding brain processes. Attention definitely falls into this category.

While some key details are yet to be discovered, the generally accepted view of how attention works in the brain involves a certain kind of battle between neurons, with the direction of our attention emerging from the outcome of this collective neuronal war. In this situation, the parallels with biological natural selection are very apparent, although the winners and losers in this particular survival of the fittest aren't organisms or even neurons— no neurons actually die in these fights. Instead, the clashes emerge a level above that of the neurons, with competing sources of information all jostling for finite attentional resources.

In order to explain this fierce neuronal competition in greater detail, I need to provide some background about how neurons interact to process information.

Journalists appear to fall over themselves to cover any result of the formula, "Scientists have discovered the brain region for X." While mapping brain regions according to function does provide useful clues to how we think, and ultimately how we might be conscious, there can be a danger in mistaking a real model of the mechanism of thought for what sometimes appears little more than stamp collecting. But if such labels aren't a sufficient explanation for how our thinking is put together, what is?

In 1961 the celebrated theoretical physicist Richard Feynman was asked to take over the introductory physics lectures at Caltech. As he was not particularly known for his interest in students, this was a slightly surprising request, but the Caltech physics lectures had become so antiquated and piece-meal, and Feynman's didactic skills so renowned, that it seemed the natural choice. The result was the most famous student physics lectures—and sub-sequent textbooks—of all time: *The Feynman Lectures on Physics*. Toward the beginning of the very first lecture, Feynman told his packed class of fresh-faced undergraduates:

> If, in some cataclysm, all of scientific knowledge were to be destroyed, and only one sentence passed on to the next generation of creatures, what statement would contain the most information in the fewest words? I believe it is the atomic hypothesis that *all things are made of atoms—little particles that move around in perpetual motion, attracting each other when they are a little distance apart, but repelling upon being squeezed into one another.* In that one sentence, you will see, there is an enormous amount of information about the world, if just a little imagination and thinking are applied.

In neuroscience, our atomic equivalent is the neuron, and the one neu-roscientific sentence we might want to pass on to that next generation of creatures, before we all perished in that cataclysm, might be: *All conscious and unconscious mental processing equates to the electrical activity of vast col-lections of neurons—information-processing brain cells, each of which has a biological version of thousands of input and output wires connected to other neurons, thus allowing each neuron to influence, and be influenced by, the ac-tivity of many others.*

A single neuron is essentially a simple node in our biological computa-tional network. It has a multitude of branches surrounding its cell body, each ending in a single wire that receives an input from another connected neu-ron. The neuron also has a long tail that splits up into many output wires, sometimes counted in the thousands, allowing it to send a signal out to an extensive selection of other neurons. The signal is usually a simple electrical firing at a standard voltage, so basically a binary code, with no firing being a 0 and firing being a 1. Almost all other neurons will be working with the same binary language. One more key fact: The wires between these neurons usually do not actually touch. Instead, when a neuron fires, it releases a small amount of a chemical known as a neurotransmitter in the gap between the wires, and it is this chemical signal that the other neuron picks up. There are many different types of such chemicals in the brain—some are designed to suppress activity, while others may enhance it.

Let's expand on the e-mail analogy of Chapter 1 to illustrate certain features of this neural system: Say I am a manager in a large company, a Fortune 500 conglomerate. Because we're so cutting-edge, we've done away with phones and rely exclusively on e-mail. As a result of this "enlightened" decision, I'm constantly deluged with e-mails. I cannot respond to every single one. So I set up a rule that if and only if I receive more than 100 e-mails on a particular topic (they *all* have the same topic of "FIRE," or 1, but we'll ignore this for the moment), then I will pass it on to everyone in my address book. It happens that I receive a lot of e-mails from one particular person, N. Uron. And I actually find myself frequently writing to him as well—strangely, he seems to be either active or twiddling his thumbs just when I am. It would therefore make sense to prioritize this closely similar employee's e-mails above the others, and if I get an e-mail from him, class it as the same as 10 other people, so that I need fewer e-mails on a topic, if he's one of the senders, before I decide to pass it on.

There are some times when I'm feeling rather less productive than others. If we all have to work nights, then I'm going to feel tired and resentful that I'm stuck in this damn office without overtime. At those times, I simply ignore half the e-mails I receive. So to get me bothered enough to forward some message, I'm going to have to receive it 200 times. And in the middle of the night, everyone else around me is exhausted, despondent, and not exactly glowing with company pride. So everyone becomes quieter at night, and messages tend to die soon after they are sent. On the other hand, in the morning, 20 minutes after the coffee break, we're all buzzing and almost begging for some activity to latch onto. At these times, just fifty messages of the same topic, on average, will make me forward the information to those around me. And because everyone else is also buzzing, much information flows around the company building, at breakneck speed.

This analogy illustrates the flexibility in a network of neurons. A neuron will only fire if it receives a certain number of firing outputs from other neurons, but this number can be increased or decreased in the short term, depending on what signaling chemicals are used in the gaps between neurons (the tiredness of night versus the caffeine fueled morning). These chemicals normally flood large collections of neurons indiscriminately. And if two neurons co-fire frequently, this in itself will physically strengthen the connections between the neurons and make it more likely that they will fire together in the future (the regular e-mails between the narrator and N. Uron). This leads to the famous neuroscience dictum that "neurons that fire together, wire together" (this is known as Hebb's law, after the pioneer of the computational study of networks of neurons). This easing of information transmission between similarly behaving connected neurons is thought to be the main microscopic mechanism for learning and memory.

If we learn, say, that the word "square" corresponds to the visual input of a square, the neurons that visually represent an object with four equal sides at right angles to each other are becoming e-mail friends with those that hear the sound "square"—they prioritize their e-mails between each other, or, in other words, the visual square population of neurons collectively strengthen their connections with the sound "square" neurons. If there is any location to this learning, it is in the enhanced sensitivity between these groups of neurons, enabling them to activate each other more easily.

When I was a first-year undergraduate, my intuition was that there may be one or a few neurons that recognized my grandmother's face, another local set that recognized a hammer, and so on. I naively saw the brain rather like a gigantic group of filing cabinets, with familiar buildings, say, neatly placed in one filing cabinet and the plots of novels I'd read alphabetized in another.

When I was told that in fact information is distributed throughout a network of neurons, that it is encoded partly in the strengths of the connections *between* neurons in huge networks spanning millions of neurons, that it isn't really localized at all, but is a pattern of activity, I found this account shocking. It is in some ways the single most difficult neuroscientific concept to accept, but because of the way evolution works, it can't be any other way. This system of neurons learning by building links with other neurons in a distributed fashion is a framework that can start small, but scale up exponentially, while some filing-cabinet version of the brain could only be made with a prospective plan by some god, not by evolution. It may be that a nematode worm, with its minuscule network of 302 neurons, can only learn a handful of things in its few weeks of existence, while I, with my 85 billion neurons, can learn many thousands of facts in my set of decades—but we nevertheless closely share the same underlying neuronal mechanisms for information processing (these worms even use many of the same neurotransmitter chemicals that we do).

This apparently simple neural system allows for incredible flexibility, especially in its larger forms. Just as DNA is written in a language understood by all life on earth, this ubiquitous neuronal binary language is potentially understood by the whole brain, which, for instance, allows for the easy exchange of information between regions. This helps explain why ferret brains can be rewired so that the auditory cortex can start to "see" if given input from the eyes, and why the visual cortex in blind people can easily adapt to process Braille.

ATTENTION AS A BRUTAL NEURONAL WAR

The semi-chaotic activity of our 85 billion neurons undergoes a kind of temporary natural selection every moment of our waking lives, as attention shapes the contents of consciousness. Rival coalitions of neurons compete with one another to be heard the loudest. Those with the most powerful voice recruit others to their cause, and suppress any dissenters, until the strongest thought is carried by millions of neurons, all with one voice—for instance, to look for the black hair of your lover as she approaches down the street. Every time you have a new thought, that idea has become the dominant clan of your internal world, following a violent battle between jostling, screaming tribes.

To illustrate just how attention shapes consciousness at the neural level, let me return to the e-mail analogy: Now I mentioned before that I'm a manager in a large company. In point of fact, I happen to work within the security department, and it's my job to see that certain unwanted types keep well clear of the office building. Each security executive specializes in a single color (there is a visual color-processing region in the human brain, known as V4, with individual neurons firing most strongly for specific colors). My color happens to be a particular shade of yellow. If the junior boys at the back of the building (primary visual cortex, the first cortical station from the eyes) send me an e-mail that they've spotted my own lovely shade of yellow on the street, then a few e-mails from them will get me jumping up and down in my seat. I want to tell everyone I know that *my* color is around, so I fire off my e-mails quick as a flash, with lots of exclamation marks at the end of each sentence. The guy in the cubicle next to me is responsible for the color black. Now if we *both* start sending e-mails madly that our colors are around, then the security guy a level above me, down the corridor, just needs one or two e-mails from us both together for all hell to break loose. You see, if he learns that both black and yellow are on the jackets of those outside from even a few e-mails, he sends high-priority e-mails to everyone he knows—and he knows some pretty senior people. (Further into the visual stream, for instance in the inferotemporal cortex in the temporal lobes, there are neurons that are responsive to combinations of features, or even specific objects, by the connections they've made with other neurons. In the neural equivalent of the current example, a wasp has been detected, and the person is allergic to them.)

Although people rarely listen to me, when the security manager responsible for yellow and black combinations sends e-mails about what he's detected, they *always* listen to him. Everyone in the entire building passes on his information without question as soon as they get it. Thanks to a flurry of focused e-mails on the topic, all the security cameras swing to the sight of the spy, everyone discusses what to do with this enemy and what it might mean, and any guards outside get ready to move the interloper along, if need be. Almost immediately, absolutely the whole building knows about the spy, with his yellow and black clothes. The guys in my office who are responsible for any colors that aren't yellow or black basically go for a break, as they know no one is going to listen to their e-mails for a while. I, for a change, become important, along with the guy sitting next to me who is responsible for black. Our e-mails are given priority—they are read first and acted on more often than not—as we report on the latest whereabouts of this probable spy to the security guy in the next corridor.

Although in the real brain, the guys responsible for yellow or black would be represented by the amalgam of thousands or even millions of similar neurons, this analogy illustrates many aspects of how attention can be generated in the brain. One important feature is that as information flows through the cortex, it increasingly gets filtered and combined to reveal its hidden, richer meaning.

Initially, our senses are constantly performing the first stage of filtering. As soon as visual information enters our cortex, any data concerning edges are already preferentially filtered for us. Then each simple detail activates a set of neurons tuned to only one or two basic features, but these neurons pass on that relatively raw information to later, more sophisticated neurons that are designed to represent ideas not just about single features, but combinations, or—even later in the stream—actual objects.

All of these separate neuronal populations, each activated by the various sensory inputs received at that moment, are trying to broadcast their own information, and each group is competing to shout with the loudest voice. But if certain combinations of basic features are spotted in the world that correspond, say, to a well-learned danger, then these set off a chain of actions in the brain. Those neurons that have detected the threat are given priority over any others.

This winning signal now recruits all related areas. If you've experienced a fear of wasps, and one is buzzing around, anything remotely wasp-like suddenly is mistaken for a wasp—you initially think that horsefly over there is one too, and perhaps even, for a moment, mistake the fridge sound for a wasp's noise. This is because all the neurons relevant to the details of a wasp are so active that these primed neurons are latching onto any tiny hint of it—and many false positive reports can occur. But if a true wasp is spotted, then all relevant neuronal regions will activate faster than usual and collaborate on recognizing the threat and then avoiding it. In the meantime, any neurons not currently coding for this danger will have their activity suppressed so as not to get in the way.

In other words, there is a constant competition occurring in the brain between different factions of neurons representing different chunks of information. Those groups that have the highest current biological relevance are given a leg up in activity, a head start, so that their influence rapidly spreads to more and more brain regions, until many areas of the brain will be dealing with this piece of information in their own way—and this amplification for this object or feature will be bouncing back and forth in the brain, constantly reinforced. At that point, the competition has effectively been won by this source of information, and any neurons representing some competing information will not only fail to broadcast their data widely, but have their activity inhibited by the neurons that *are* dealing with the relevant information. The winners in these battles are indeed oppressive victors, squashing any potential dissent, but in this way attention remains focused and we can respond efficiently to any danger, without being distracted.

This ideas-based, winner-takes-all system sounds simple, but emerging out of it is an amazingly flexible mechanism for much of the cortex in concert to shape itself according to some current purpose, whether it be a biological threat, such as a wasp sting, or a more complex task we've consciously set, such as composing a piece of writing. Large swaths of the cortex can recreate its own mountainous landscape of combinations of activated or inhibited neuronal coalitions in a highly pointed way, reflecting all sensory, memory, cognitive, and motor features of the goal of the moment.

ATTENTIONAL VICTORIES EMERGING INTO CONSCIOUSNESS

At what point does this attentional battle turn into consciousness? In line with Libet's free-will experiments and Nikolov's neuronal decision modeling data, at the first ramping up of decision activity, and the onset of the attentional war, there is no sign of awareness. Instead, neuronal activity probably becomes conscious when the battle is clearly won, when all features of a goal are significantly bound together in frenetic activity throughout the cortex, and all nonrelevant details are simultaneously shut down.

Consequently, when we spot that wasp, we don't see lines of yellow independent from lines of black, both distinct from a couple of wings and all quite separate from a vague buzzing sound. Instead, the collaborative endpoint of the processing bursts through into awareness. We immediately know we are near a wasp, can hear and see it very much as a single yet compound object, orient our heads toward it, and already are thinking where to move away from it or what means we can use to push it out the window—and little else aside from that wasp at that moment occupies our consciousness.

The process of combining more primitive pieces of information to create something more meaningful is a crucial aspect both of learning and of consciousness and is one of the defining features of human experience. Once we have reached adulthood, we have decades of intensive learning behind us, where the discovery of thousands of useful combinations of features, as well as combinations of combinations and so on, has collectively generated an amazingly rich, hierarchical model of the world. Inside us is also written a multitude of mini strategies about how to direct our attention in order to maximize further learning. We can allow our attention to roam anywhere around us and glean interesting new clues about any facet of our local environment, to compare with and potentially add to our extensive internal model.

The example of our attentional system being driven by some biologically important external object may be the normal form of attention in the animal kingdom, where most species have a considerably simpler mental life than we do. But if you have far greater processing capacity and a more elaborate internal model of the world, then you also have far more choice about what to attend to, as so much more than the latest obvious threat or sign of food could potentially aid your biological goals, if analyzed carefully.

Without any obvious external threat forcing itself on our attentional system, how is the choice made to attend to any of the seemingly infinite options? On the one hand, consciousness and complex thought can constrain the process in a number of ways. We can logically interrogate the reasoning behind each major option and heavily favor one side in the neuronal battle to follow. We are certainly not limited to the small set of more instinctive attentional filtering systems. Instead, we can consciously create almost any kind of neuronal filter, strongly boosting attention for one feature of the inner or outer world and suppressing others. In this way, we can seemingly choose what to attend to—in other words, what to be aware of.

On the other hand, despite our impressive conscious ability to bias what we attend to, the basic competitive neuronal mechanism of attention is just the same in these seemingly voluntary, internal choices of attention as it is for the immediate external drivers of our attentional system. Out of the many fighting voices in our minds, conscious control is but one choice, commonly pitted against a set of various bullying unconscious desires, where only a single voice can win out, to ruthlessly recruit to its cause all conceptually related neurons throughout the cortex and suppress any dissenters.

Here, with these neuronal wars, we can return to the question of emergentism, where advanced ideas on one level emerge out of the interactions of simpler, lower-level objects, with a clear proposal for the mechanism of emergentism in relation to awareness: Multiple factions of neurons competitively interact, with two kinds of feedback—a positive form that can rapidly boost neuronal activity, and a negative form that can rapidly inhibit it. The complex interplay between these two opposing feedback loops at the level of local neurons can dynamically tune and activate much of the cortex, giving rise to highly flexible, global, synergistic information processing and consciousness.*

Overestimating the Value of Emotions

So far in this chapter, I've been describing how attention is a key component of consciousness. Attention is a filtering and boosting mechanism, taking the entirety of the sensory input we receive, including much that is irrelevant to us, and converting this into a far more finite, refined output containing only those items that are most germane to our current goals. It is this output that we are conscious of and that I will explore in more detail for the remainder of this chapter. But first, I turn to two particular forms of conscious content: emotions and self-consciousness.

It would be churlish to underestimate the role that emotions play in our conscious lives. Emotions continuously, profoundly shape our thoughts and behavior, and many would say it is the panoply of so many vivid sentiments that give true color and meaning to life. I don't deny any of this. But some theorists have put emotions as both the core evolutionary driving force of consciousness and its main contents. There are good reasons to reject such a position.

For instance, there are many moments in our lives when we're not conscious of any particular emotion, even though we're nevertheless very much aware of something.

But what would happen if we lived our entire lives in such a grey state? Would we still be conscious? In the book *Descartes' Error*, Antonio Damasio discussed a patient, known as Elliot, who in many ways is a modern version of Phineas Gage. Elliot, also following damage to his orbitofrontal cortex, also underwent a radically shifted mental world. His life is now almost devoid of any emotions whatsoever. He goes through life continuously in a

---

*It's possible that in most, if not all, other examples of emergentism one would care to mention—for example, the collective intelligence of ants or macroeconomics—the recipe is little more than positive and negative feedback loops at the lower levels interacting to generate a more accurate, fitting informational solution at the level above.

neutral emotional gear. But his level of consciousness seems hardly dented, even if his social decisions are abnormal, just as Gage's were. In total contrast, a severe attentional deficit following brain damage, as I'll describe in the next chapter, crushes awareness.

At the same time, intense emotions can seem to shrink consciousness in unhealthy ways. For instance, some people become sufficiently nervous when speaking in public that they stumble over words, or momentarily forget what they meant to say. Their minds go blank, and it almost feels as if their awareness of everything except the object of their nerves has disappeared. Ramp this up many notches, and terror, an emotion designed to save our lives, regularly kills because it profoundly diminishes consciousness and therefore intelligent control. For instance, it is tragically common in plane crashes for passengers to be trapped in their seats because they repeatedly press the safety belt as if it had a car's release button instead of the plane's metal lever. The most minuscule conscious analysis of the situation is unavailable to them because they are paralyzed by fear.

LAYERS OF AWARENESS

A closely related topic is self-awareness. Some theorists claim that we are only truly conscious when we are self-aware, and that our sense of self is the most critical component of any conscious experience. Again it is suggested that evolutionary pressures for a more sophisticated model of oneself were responsible for generating consciousness in the first place.

Self-awareness is actually a rather confusing term. It has at least three different meanings. The first, less grandiose definition of self-awareness is simply the state in which an animal is aware of itself as distinct from all the animate and inanimate objects around it, and it thinks and acts in accord with this basic assumption. In a sense, all creatures need to make this conceptual distinction, irrespective of consciousness, so any creature that also happens to be conscious will automatically have this version of self-awareness. This may seem a trivial point, but it emphasizes how embedded within our biological makeup this heavy distinction between us and the rest of the universe is. Self-awareness based on this definition may not be a necessary component of consciousness, but instead just an accidental consequence of something being a conscious animal with an evolutionary heritage.

Indeed, the case of the conjoined twins Tatiana and Krista, who appear to have conscious experiences that are not their own, provide tentative evidence that this accidental combination need not always occur. Occasionally, under special circumstances like this, we might be able to have experiences, but not be the owner of those experiences.

Various psychiatric populations also hint at a possible separation between consciousness and the self. For instance, in certain cases of multiple personality disorder, patients may attribute many of their experiences to other personalities within them. One patient reported "Joy is happy and playful, so sometimes when I'm down she becomes me. Sometimes it cheers me up, but sometimes it is only Joy who is happy and I'm still upset." Here, seemingly as a strategy to improve her mood, this patient lets an alternative identity take over her experiences.

Although multiple personality disorder is a controversial diagnosis (it has been suggested that all such patients fabricate these extra personalities in a desperate attempt to protect them from some past trauma), less controversial is an analogous situation in schizophrenia. One of the hallmarks of schizophrenia is a genuine belief that the voices within your own mind are not your own. In other words, many schizophrenics are convinced that some part of their own experiences at least partially belongs to someone else.

Although rather circumstantial, all these pieces of evidence point to a potential loosening of the glue between experience and a sense of self, reaffirming the possibility that their apparent inseparability might be accidental.

But many people use the term "self-awareness" in rather more abstract ways. One version involves being aware of ourselves as having this particular body, this particular face, this persona, and so on. The main test for this form of self-awareness is whether you can recognize yourself in the mirror. But one critical question following from this definition is whether self-awareness is the cause of extensive consciousness, or simply a consequence of it.

For an animal to be able to know that this other animal in the mirror is in fact itself is a tremendously mentally demanding feat. All the animals it has met in its life so far have been other animals, so there is a strong, very natural expectation that this animal in the mirror is another animal as well. For the animal to understand that the reflection is itself, it needs to acknowledge the majority of the following cues: It needs to realize that the other animal's touch is incongruous, being cold and hard; that the other animal is missing any scent; that the rest of the room in the reflected world is an identical copy of its own room; that every time it makes a movement, the other animal copies it perfectly; and that no other animal in the world could match its own actions so quickly.

In fact, it gets even more complicated in the lab, where, in addition to all the above, the animal needs to perform a secondary task to prove beyond doubt that it recognizes itself in the mirror: usually a dye is placed on the animal's head in a location that it can't normally see, but is easy to spot in a mirror—the forehead is a popular place. For the researcher to gain proof of self-awareness, the animal has to recognize the mirror animal as itself, and it further needs to realize that this spot is a new unnatural addition to its facial makeup, before finally having the motor ability to touch the colored spot on its own body.

Being a shameless cognitive neuroscientist at heart, I've subjected my own baby daughter to this task multiple times as she's developed, and also regularly encouraged her to play in front of a mirror more generally. As a parent, I had a sense early on that she recognized herself. For instance, for many months, while she was rather wary of other babies, she'd be fearless of the baby in the mirror, approaching it with glee. But this wasn't proof. Solid evidence of the above kind only came when she was nearly fourteen months old, and she tried to remove the new streak of color on her forehead, via the mirror. It was clear from her less than usually fluid movements that using the visual feedback from the mirror, rather than from her own kinesthetic and more direct visual senses, was a distinctly unnatural and difficult feat in itself, over and above all the other very complex requirements of mirror self-recognition.

Obviously as a diligent scientist I made sure I could repeat the event multiple times! It was also clear on the first couple of goes that she noticed the change, as she laughed as she looked at the spot, but it didn't automatically follow that she wanted to remove it—she might even have liked her new facial feature! For the first few trials, only when I was able to cheat and use language, asking her to remove the object on her face, did she actually do it. But then in later trials it became a fun game to use the mirror to remove the color, and I no longer needed to prompt her. This clearly demonstrated to me just how many different ways an animal could fail at the mirror recognition test, even if it clearly had the ability to pass.

So there is a great multitude of hidden, complex assumptions required to demonstrate that you can recognize yourself in the mirror. Only a considerable intelligence, and a high level of consciousness, with motivation directed in the right way, would be able to pass such a test. Therefore, it seems likely that self-awareness in this sense is a side effect of a powerful intellect and rich conscious life, rather than the cause of either of these.

The final, to my mind most intriguing, version of self-awareness is where you are aware of your own consciousness. For instance, I might watch my baby daughter sleeping deeply, and not only experience feelings of love and pride, but also become aware that I'm having these emotions and say to myself, "Oh look—right now I'm experiencing feelings of love and pride." It's assumed that whenever we use language to communicate our feelings and sensations, we are relying on self-awareness, since we have to probe our own experiences, as if on a higher plane of consciousness, in order to know what they are. Many theorists, especially of a philosophical bent, believe that this "higher order" consciousness is the only kind of consciousness that really matters.

The theory comes in a multitude of somewhat confusing flavors, but before I discuss it directly, I would first like to digress in order to show that, for certain forms of content, and with a relatively standard reading of the theory, we have this form of awareness far less than we think we do. In fact, there is a striking contrast between our level of consciousness, which is undoubtedly incredibly rich and varied, and our level of insight into our own conscious minds, which is patchy and deeply unreliable.

We are all in some ways appalling at making decisions, because our more primitive drives heavily bias us to think and act in short-term ways, so that we can survive and reproduce now, today, regardless of the next month, year, or decade. Some of us can also easily lose control in blinding waves of rage or jealousy, and more generally act based on some emotion, but at the same time be rather oblivious to the emotion. For instance, we may be so focused on the object of our rage that we have no spare attention to stop and notice that we are actually angry.

A lack of insight into one's own feelings and motivations is also a key trait in almost all mental illnesses. For instance, a schizophrenic doesn't understand that he is delusional, and a depressive may not realize that she is feeling irritable or subdued until a full depressive episode has completely taken her.

This is certainly not a pattern of a continuous, proficient facility for self-awareness, at least where emotions are concerned. Instead, we seem to spend surprisingly large amounts of time being unaware of the emotions we are currently experiencing, presumably because we don't attend to them. After all, emotions are there to guide us toward or away from certain objects or activities, so it is natural to attend to these objects while ignoring our own feelings. It is distinctly unnatural (although often extremely useful) to pause and attend to the emotion itself and the quality of reasoning behind it.

But what of our senses? Do we also lack insight into our own perceptions? One recent study that helps answer this question involved subjects viewing a set of six striped circles, and then a virtually identical second set. Within either the first or second group of circles was a single striped circle that was a little more vividly striped than the others. Subjects first had to guess which of the two sets had this odd-one-out stimulus, and then they had to rate their confidence in this guess. As the trials went on, the experimental program was continuously changing the detectability of the odd-one-out stimulus based on the subject's performance, so that accuracy was maintained at close to 71 percent for all participants—therefore well above chance, but still an extremely difficult task. This fixing of performance meant that the only factor that could change between subjects was how well their confidence in their decisions mapped onto their accuracy. So subjects adept at being aware of their own perceptions would almost always rate their confidence as very high when they were correct in spotting the odd-one-out feature and rate their confidence as low if they were wrong. One striking result of this study was that although some subjects were indeed quite proficient at this task, there was a huge variation, and many subjects were very poor at matching their confidence to their accuracy, regularly either being highly confident that they had guessed right when they were wrong or having no confidence in a correct decision.

When I pored over the details of this study, the first thing I asked Steve Fleming, its main author, was whether the subjects who were weak at reading their own minds were otherwise impaired in any way. It was an obvious thing to look for, and he'd already carried out an extensive analysis on this, but couldn't seem to find anything at all. Those at the bottom range were just as good at basic perceptual tasks, seemed just as bright, and in all other ways seemed just like the rest of the group, except for having poor insight into their own perceptions.

Therefore, although we clearly also have the capacity to be aware of the contents of our minds in this higher-order way, that certainly doesn't mean we're all fantastic at it. In fact, some of us, who are otherwise quite normal and apparently just as conscious of the world, are very poor at it indeed.

This observation of the patchiness of self-awareness, in this higher-order way, needn't be an outright attack on the theory. Although it really does feel to us that we spend every moment of our waking lives conscious, it's possible that we're utterly mistaken, and there's only a far more limited set of moments when we are truly conscious in this higher-order sense.

Although it might appear unintuitive, but technically possible, that we're mistaken when we assume we're conscious whenever we're awake, another feature of the theory seems far more troubling: If I silently stare at the blank wall beside me, with a quiet mind, I'm clearly conscious of the wall, and yet I do not seem to be having any thoughts about my perception. A higher-order consciousness proponent would claim that I necessarily have some higher-level thought or perception about my basic perception of the wall, because I must in order to be conscious of it, even if I don't realize that this process is going on. But then the theory is in danger of being circular or otherwise empty, and it's unclear how you could ever verify or falsify such a position with experiments.

Another issue for the theory is its approach to the utility of consciousness. The leading defender of this theory, David Rosenthal, believes that one consequence of viewing consciousness in this higher-order way is that there is nothing useful or advantageous about being conscious, since our cognitive skills occur at a level below that of consciousness, which observes knowledge acquisition passively from above, as it were. So awareness serves no evolutionary purpose and provides no enhancement to the quality of our learning. There is overwhelming evidence against this position: Consciousness clearly is necessary for any form of complex learning to occur, which in itself is a good reason to reject this higher-order theory of consciousness.

As far as I know, any detailed discussion of a mechanism for consciousness within this theory simply stops at the suggestion that reflexive cognition—having a thought of a perception, say—is how consciousness comes about. There is virtually no description of what thought or perception means here from the context of standard psychological components or brain processes. There is virtually no detail about how the bridge between higher and lower mental levels might work. There is little or no explanation for how or why full consciousness should be equated with this reflexive step, how it fits in with information processing, what the evolutionary basis for higher-order awareness is, or what the overall purpose of such a conception of consciousness could be.

A more scientific approach, as I'm describing in these pages, is potentially far more detailed and profound. For instance, although I will describe other components in a moment, I've so far outlined in this chapter how attention is one key component of consciousness. Attention is a well-studied process, both psychologically and biologically. It immediately casts doubt on the necessity of some higher-order level for consciousness, and instead suggests that, at least in some situations, consciousness might emerge from the winner-takes-all neuronal battles that occur unconsciously. Attention puts information processing at the heart of consciousness and suggests that consciousness is the end product of an aggressive data-filtering and -boosting process.

So self-awareness, in any of its guises, appears to be a side product of both a deep intellect and a rich conscious life, rather than a cause of our extensive awareness. Instead, both emotions and, more generally, any information one has about oneself are only special for the biological importance they carry in keeping us alive. From the perspective of consciousness, they are just another kind of information we could be aware of, out of the millions of possible experiences we could have, many having little to do with either our feelings or our sense of self.

But although I believe that theories defending the primacy of self-awareness, particularly involving higher orders of consciousness, are unhelpful ways of looking at the problem, there is one feature of these positions to which I am very sympathetic. Being aware of oneself, or of one's own thoughts and sensations, might be an accidental side product of a burgeoning consciousness, but it is nevertheless a profound side product. Such examples join a much wider group of important conscious events that are highly conceptual, sitting at the very top of a mental pyramid of ideas. This general layering of concepts, with consciousness at the top, allows us to experience our surroundings not as a bland sheet of raw data but as a vibrant, immensely patterned picture, utterly pregnant with meaning, which allows us to glide through this landscape with exquisite, effortless control.

For the remainder of this chapter, as I move from the mechanism that chooses what content to populate consciousness to the contents of awareness themselves, I'll be repeatedly highlighting the importance to consciousness of building and manipulating these intricate monoliths of knowledge.

Four Compartments to Awareness and No More . . .

Although so far I've talked about how attention acts as a filtering and boosting mechanism, I haven't yet shown just how aggressively the brain can filter its input, or how intensely it can boost certain input signals. In fact, routinely, attention filters the billions of pieces of information streaming into our senses, or bouncing around our unconscious minds, into a maximum of three or four conscious items. So the filtering process is about as aggressive as one could imagine. But the boosting process can compensate for this limitation just as aggressively: Each of the mere handful of items can be an immensely complex mental object, and although their number is painfully finite, these conscious objects can be assessed, compared, and manipulated in virtually any way imaginable.

This tiny, yet ever so powerful output store of attention is our "working memory." Working memory is an inherently conscious short-term memory container where we can remember, rearrange, and evaluate whatever is in this group of items, even if it comes from different senses or categories.

Over the past twenty years, the most prevalent, popular psychological theory of consciousness has been the "global workspace theory" proposed by Bernard Baars. In many ways, Baars' ideas resemble mainstream views on the psychology of attention. In the global workspace theory, there is again an unconscious fight for dominance between low-level coalitions of neurons, with a winner-takes-all attitude. The winner filters into consciousness, where again Baars makes more parallels with attention, by talking of a spotlight directed onto only a small portion of a theater stage. This spotlight is the subset of our world that we are actually conscious of, and it broadcasts itself to the whole audience, in other words, making just a small number of items available to much of the brain, potentially for further information combination and comparison. But Baars' most bold and interesting claim is that, more or less, consciousness boils down to the information sitting right now in our working memory. He views working memory as existing for a second or two, available to almost every corner of the brain, and there to guide unconscious specialized knowledge regions to help us carry out our most complex tasks, such as language and planning.

Although there are ambiguities in the definition of "working memory," in the main I firmly agree with Baars that consciousness and working memory are largely synonymous processes, and that attention is the critical means by which items enter into consciousness. But the next key step, from the point of view of consciousness, is to fill in the details, to describe exactly how working memory functions, both psychologically and in the brain. Twenty years after Baars first formulated his global workspace theory, our understanding of working memory and attention is now far more comprehensive. And, with these advances, many mysteries of consciousness are being solved.

The first feature of working memory is how it is surprisingly so limited in capacity, comprising a mere handful of conscious objects. Many different experiments have confirmed this constraint on our conscious space—though each study has had to take careful precautions to counteract our prodigious ability to develop strategies to cheat—to try to enhance our capacity, usually by linking current items to our long-term memory store. The standard methods for removing the opportunity for such strategies are either to present stimuli so briefly that our myriad workaround tricks don't have a chance to form, or to present more abstract items that have absolutely no relation to our preexisting memory.

For instance, in one landmark early study, George Sperling presented subjects with a grid of 12 letters, in 3 rows of 4, but only for about 50 milliseconds. Subjects then had to report as many of the letters as possible. They would get, on average, about 1.3 letters per row correct, or about 4 items in total (1.3 multiplied by the 3 rows is 3.9). In a fascinating twist, in some trials, Sperling also immediately followed the flash of letters with a cue to tell the subjects to give their answers from just a single particular row. Now, very surprisingly, they would generally correctly recall all 4 letters from one row instead of the 1.3 letters per row that they could previously manage, presumably because the immediate instruction enabled their attentional system to focus fully in on this one row before the fresh visual information faded. If instead the cue to center on a single row came a second or more after the grid had disappeared, then subjects returned to their previous performance, as if no cue had occurred, and could only answer about 1.3 items from this cued row. Within this single second, their attentional system, not knowing which row to focus on, had applied equal importance to all 12 items as they all faded from their initial fresh visual state, and only the letters from 4 random locations in the entire grid could be preserved in their limited short-term memory store.

A conscious limit of 4 objects turns up faithfully in almost any kind of experiment one tries. But in real life we do not usually need to remember letters in a grid, so I'll share another example that will seem more natural. We commonly track multiple moving objects—maybe a group of people on the street that we walk past, or a set of players on a soccer pitch. Animals in the wild may also need to analyze where a group of other objects are moving. For instance, the members of a chimpanzee tribe may need to monitor the location of each member of a competing tribe that is encroaching on their territory. In an experiment that mirrors these everyday skills, Steven Yantis presented subjects with a set of 10 crosses on the computer screen. A subset of these initially flashed, and subjects had to keep track of them as they moved randomly around the screen and ignore the moving crosses that previously hadn't flashed. At some point, the moving crosses would become stationary, and subjects had to say which of the crosses were the initially flashing ones. If there were only 3 crosses to keep track of, then subjects found this task relatively easy. When volunteers had to simultaneously track 4 objects, they were somewhat less accurate, but still performed the task competently. When Yantis increased the number for the volunteers by 1, to 5 moving crosses to keep track of around the screen, because this number exceeded their working-memory capacity by a single item, most subjects found this variant of the task virtually impossible. This experiment is a striking demonstration of how sharp a barrier this capacity of 4 conscious items is.

Surprisingly, our working memory limit of a handful of items is basically the same as the monkey's, even though a monkey brain is about one-fifteenth the size of ours. And our closely related skill of being able to recognize the number of items briefly presented to us—about 3 or 4 again, before we need to start approximating—is the same capacity limit that newborns have. In

fact, many other species have the same upper bound to immediately counting the number of objects, including the lowly honeybee, which can differ-
entiate patterns containing 2 from 3 items, or 3 from 4, but not 4 from 5 or
above. So there may be something fundamentally limited about just how
many items all animals can store in short-term memory.*

...But Each Conscious Compartment Can Hold Objects of Great Complexity

That the contents of consciousness, if you discount compensating strategies,
is fixed at about four items seems to be a tremendous handicap. But in humans, especially, *one should never discount strategies*. We use built-in attentional mechanisms as well as the heavy ammunition of our conscious powers
of analysis to regularly load huge quantities of data into each conscious compartment, shamelessly cheating our apparent working memory boundaries.

Turning first to the role that attention plays in boosting our capacity per
working memory holder: Once attention has decided to prioritize a given
object, whatever it may be, the neuronal war has been won. Activity in much
of the brain is then shaped according to this current object and how it relates

---

*The cause of this surprising limitation is a matter of intense current debate. One expla-
nation suggests that three or four items are plenty, in almost all circumstances, to be the focus
of our attention. There are exceptions, however. A student who is struggling with a mathematics assignment at school, or someone trying to learn the ins and outs of the latest overly
complex gadget, may well wish it were possible for the human mind to handle more. But in
our evolutionary past, we were rarely simultaneously faced with more than a few predators
about to eat us, and we rarely chased after more than one or two sources of food or potential
sexual partners. We didn't need more in order to face the dangers of the world and take advantage of its benefits. In other words, perhaps there was never an evolutionary need to attend
to more than just a few items at a time.

Another explanation, supported by computational models of how neurons interact, suggests that somewhere between three and four items is actually the maximum that can be practically sustained within a brain. Once attention has increased the signal for a given item, any
neurons throughout the cortex that relate to any features of the object not only fire more actively, but also link together in a signature rhythm. If more than one object is being attended
to simultaneously, multiple brain rhythms are required in order to keep the signals separate.
But the brain can only sustain about three or four of these harmonies—any more and they
begin to blend into each other too much, and the result becomes a disjunctive neural noise,
where objects become confused with each other or simply forgotten.

to us. For basic objects or features in the world, such as the color red as painted on a plain wall, attention boosts the signal by enhancing the readiness to fire of our visual regions, especially those for red. Non-red color-coding neurons may be suppressed, not only in our color-processing centers, but everywhere else as well. Our hearing and taste centers, for instance, may be inhibited. At the same time, all general-purpose regions, especially the prefrontal and parietal cortices, which are closely connected to consciousness, have activity that hones in on this current feature. All of this works well, and does help us spot red in the world, but the effects are not nearly as striking as when the brain has some internal hook by which to latch onto, so as to enhance the incoming signal.

If, instead of the red wall, the current object of attention is Angelina Jolie on the big screen in front of me wearing a red dress, then anything around me that's not Angelina Jolie gets suppressed, and any corner of my brain with any relevant information about Angelina Jolie becomes activated. As soon as I see her, I recognize the features of her face, I know her name, recall how she speaks, have knowledge of her famous husband that I can easily retrieve, remember the other films she's been in, and so on. And, of course, I can also see that she's wearing red. These aren't sets of unrelated facts; they are all bound together as a single, unified, complex object. The previous example of the plain wall as an attended single object effectively had red as the only feature. When I attend to Angelina Jolie, the same piece of information, red, is attached to my conscious representation of her, but this time red is only one of dozens of features connected to this single mental object. This is a fantastic system to have—attention takes this raw input and seamlessly transforms it into a panoply of interconnected facts by the time it reaches consciousness. And yet, because attention has activated and drawn together all the components of this one object, Angelina Jolie, it takes up the same single compartment in my working memory as does the plain red wall.

In other words, we may only have a few conscious compartments, but each holder can cope equally well with the simplest of objects or the most complex. And the term "working memory objects" in this context generally means just some bound collection of information. It could be a physical object, like Angelina Jolie. But it could equally mean one strand of the plan I devised for this current chapter as I was walking to Grantchester.

Just how much information can one working memory object support? This is where the concept of "chunking" returns in force. In terms of grand purpose, chunking can be seen as a similar mechanism to attention: Both processes are concerned with compressing an unwieldy dataset into those small nuggets of meaning that are particularly salient. But while chunking is a marvelous complement to attention, chunking diverges from its counterpart in focusing on the compression of conscious data according to its inherent structure or the way it relates to our preexisting memories.

One of the most dramatic experiments to demonstrate how chunking can expand what we store in working memory was published in 1980 by K. Anders Ericsson and colleagues. The experiment is beautifully simple: The scientists took one normal undergraduate, with an average memory capacity and IQ for a student, and gave him a basic task—the experimenter read to him a sequence of random digits and he then had to try to say back the digits he'd heard, in the order he'd heard them—just like trying to remember a phone number someone has just said to you. If he recalled the digit sequence correctly, the next trial would be one number longer. If he said it back with any mistakes, the next trial would be one number shorter. This is a very standard test for verbal working memory. However, in this case, there was a big twist—he did this task for an hour a day, for roughly 4 days a week, *for nearly two years!*

At the start, he was able to remember about 7 numbers in a sequence, which is indeed about average (almost everyone improves on their initial verbal working memory limit of 4 through various rehearsal strategies). But as psychology experiments go, this must have potentially won a prize for the most boring in the world, being the same day in, day out, for months on end. In order to spice things up for himself, the participant seemed determined to improve his performance. And improve he did, until, by the end of the experiment, 20 months later, he could successfully say back a novel sequence that was 80 digits long! In other words, if 7 friends in turn rapidly told him their phone numbers, he could calmly wait until the last digit was spoken and then, from memory, key all 7 friends' numbers into his phone's contact list without error.

On occasion, he was tested after a session to see if he could still recall any of the sequences from earlier on in that session. At the start of the experiment, he was understandably useless, hardly remembering anything of the digit sequences, even though they were only 7 digits long. However, toward the end of the experiment nearly two years later, despite the sequences now being over 10 times longer than when he began the experiment, he could remember the vast majority of the sequences perfectly. So not only could he have immediately recalled 7 combined phone numbers, just after hearing them, but he could also have typed them in without error *an hour later*! How did he achieve this seemingly superhuman improvement in performance?

This volunteer happened to be a keen track runner, and so his first thought was to see certain number groups as running times, for instance, 3492 would be transformed into 3 minutes and 49.2 seconds, around the world-record time for running the mile. In other words, he was using his memory of well-known number sequences in athletics to prop up his working memory. This strategy worked very well, and he rapidly more than doubled his working memory capacity to nearly 20 digits. The next breakthrough some months later occurred when he realized he could combine each running time into a superstructure of 3 or 4 running times—and then group these superstructures together again. Interestingly, the number of holders he used never went above his initial capacity of just a handful of items. He just learned to cram more and more into each item in a pyramidal way, with digits linked together in 3s or 4s, and then those triplets or quadruplets of digits linked together as well in groups of 3, and so on. One item-space, one object in working memory, started holding a single digit, but after 20 months of practice, could contain as much as 24 digits.*

So, when pushed by challenging tasks, we can use our long-term memory as a crutch to convert the items in working memory into a more efficient form. The task becomes dramatically easier, our performance increases

---

*At the time of this study, a normal person using a strategy to vastly improve his working memory was a highly unusual result. But about a decade later, in the early 1990s, the world memory championships started, where many other normal people would use their own heavily practiced strategies to compete on similar tasks. Now, after about twenty years of vibrant competitions, this initial feat of remembering 80 digits seems unimpressive. Many new techniques have been explored and there has been an increasing number of serious mental athletes. The current world record holder on virtually the exact same task as just described can correctly recall 240 numbers just spoken to him in sequence. The results from other tasks are equally awe-inspiring. The current world record for numbers of cards memorized in sequence in an hour is 1,456, and the shortest time to memorize a single pack of 52 cards is 21.9 seconds!

markedly, and the newly chunked information we store is more stable, robust, and efficient.

But in humans, especially, it's not just mnemonic tricks and familiarity that can profoundly increase the actual information stored in our working memory. In the example above, the student improved his performance by artificially gluing these novel numbers to his preexisting structured knowledge about running times. He, in effect, forced patterns into unpatterned data. But often there really is a clear structure or pattern to the information streaming in from our senses, and in these situations our consciousness seems particularly alert to its detection—probably because such novel information promises significant improvements—and we can rapidly exploit this newfound knowledge.

There is good experimental evidence that we spot and successfully use any structure in sequences to aid working memory. For instance, sticking with the task of simply remembering sequences of digits, colleagues and I in Cambridge presented volunteers with novel sequences of four double digits, some of which had a hidden mathematical relationship between them, such as 49, 60, 71, 82 (so, increasing by 11 each time). Other sequences had a random spacing between items. As you'd expect, participants were considerably better at recalling the structured sequences than the random ones. Volunteers noticed the structure and found that the task became easier when there were discernible patterns, as if there had been fewer items to remember, precisely because they had found the rule that linked the digits together. Although we didn't test this, we could have given subjects patterned sequences 300 digits long, and they probably still would have had no trouble recalling the sequences—all they would have needed to remember would have been the first number, the last, and the rule. In contrast, they would have utterly floundered with 300-digit-long random sequences on their first session, and even on their 200th session. Importantly, chunking by rules is usually far more effective than chunking by memory alone.

One classic area of expertise in working memory is that of chess. We novices may look at a board full of about thirty chess pieces in some complex position and be lucky if we remember a few of those pieces. Chess masters, however, can remember almost the whole board with just a look. How do they do this? Very probably they are using a combination of memory (say, remembering the position of the pawns because pawn structures tend to be diagonal), and logic (such as perceiving doubled rooks as a powerful push up a line toward the opposing king). Chess expertise is a good illustration of how memory, logic, and strategy can sometimes inextricably intertwine, with structured information at the core.

Indeed, with the contents of working memory allowed to be virtually anything, this conscious playground of ideas is at its most powerful when the contents themselves are goals or strategies: If each mental trick can be treated as a separate building block in consciousness, where it can be combined with others in order to generate novel, more potent strategies, an unrivaled potential for learning and understanding is unleashed.

So we may be biologically constrained to consciously store only a handful of items for a few seconds, but we are also able to use any trick in the book to dramatically increase the amount of information per item. This may involve employing relatively trivial tactics, such as repeating numbers to ourselves. But we might just as easily use grander strategies, such as linking large amounts of novel information to preestablished memory chunks, or noticing the logical rule that binds many unfamiliar items together into a more coherent single unit.

BELITTLING THE RICHNESS OF EXPERIENCES?

This is an appropriate point at which to pause and meet one obvious objection to the thesis that consciousness boils down to an attention-gated working memory, with up to four chunks making up its contents. If our consciousness is really limited to a small handful of highly processed items, then how can we at least appear to see many more objects at once? It certainly seems that if I gaze up at the sky, I can make out more than four objects—maybe hundreds more in one go, and I can see them all clearly.

But I would argue that in this situation, attention is spread wide and thin, like an overblown balloon. Its thinness means that we are indeed aware of these hundreds of objects, but in a minimal, approximate way. Gazing up at the sky without any knowledge of star charts is akin to seeing the whole collection of stars as one fuzzy, complex object. If we want to remember things better, if we want to start seeing groups of stars, and memorize their relationship to each other, then—guess what?—we develop chunks to help us. I recognize the Plough because it looks like a deep frying pan with a wonky handle. And that constellation is Leo, a proud lion resting on the savannah. Without these chunks of stars, linked to well-known objects in memory, we'd have struggled to recognize any astral features, and historically this would have been a disaster for both navigation and agriculture.

There is strong experimental confirmation of this sense that the more we see, the less we actually take in of each object. For instance, if we have to identify varying numbers of letters or digits, which flash briefly on the screen, then the greater the number, the less likely we are to identify each one. In fact, however many there are, we're unlikely ever to remember more than four of them, even if we do get a vague sense of the approximate number of objects we're seeing. And, as I've discussed, in whatever way we are looking at the world, with any sense or stimuli, we only ever are fully aware of about four objects. Any impression that we are aware of more items may simply be an illusion. This illusion is partly a product of our extreme readiness to group items together to take up a single working memory slot—for instance, the *collection* of all the visible stars that we are currently viewing.

Without grouping together or fully processing any items in very busy surroundings, we usually only have a vague, faint conscious impression of its details. For instance, occasionally we may catch only the briefest glimpse of a scene and just have a gist of the objects in it. Our sense of gist reflects the fact that attention has two clear stages: The first, meager, less interesting stage of attention, and indeed awareness, is where we get a weak sense of everything around us, as if we're not really attending to anything—or rather, we are attending to everything in the same minimal way. This lasts for about 200 milliseconds. A short time later, though, the second form of attention kicks in, which is goal driven. Our neuronal landscape shapes itself according to the task at hand, and we start to hone in on interesting details—there's my wife in the station, say.

During this second stage, our brains then calculate exactly what it is we want to focus on, what the few objects are that really matter. This important subset of our world gets a generous attentional boost, and we are far more aware of what matters. Everything else gets suppressed, and our awareness of whatever is outside our working memory and focus of attention may become invisible.

Scientists have seen the neural equivalent of this story firsthand in a part of the monkey brain that codes for visual objects. Known as the "inferotemporal cortex," this region has certain neurons that fire strongly for one particular item—say, a flower—and weakly for another—maybe a mug. Leonardo Chelazzi and colleagues used electrodes to study these neurons, one by one, when a monkey was looking for a particular target object that would appear on the screen—such as the flower—to get a reward. Whether or not a given inferotemporal cortex neuron was responsive to flowers, it would initially peak in the same way, as if everything that the monkey was looking at was always provisionally interesting. Only after a few hundred milliseconds would the neuron show its true form. If the neuron wasn't interested in flowers, its activity would die away, but if it was interested, then its activity would continue to climb strongly. So these neurons, at the business end of how the brain attentionally responds to stimuli, have a two-stage firing pattern—the first is like getting a faint gist of the scene, while the second is all about carrying out a goal, with neurons shaping their activity to reflect what's important and what's not.

In our sense of gist, there is no second stage, or rather, because the input was so weak and transitory, the second stage is merely a copy of the first stage, and so there is minimal or random shaping of the input. Consciousness is unfocused and can randomly, weakly recognize a handful of objects from that brief glance.

When we grasp the gist of a scene, imperfect as it is, nevertheless at least some attention has to be involved. We can show this by seeing what happens when we completely remove attention from that brief glance. Michael Cohen and colleagues recently carried out just such a task. Subjects watched a rapid stream of different images, which changed every 100 milliseconds. Most of the scenes were just boring color swatches, but one of them in the middle was an interesting real-life scene, say of a city view replete with many skyscrapers. If this was all the subjects had to do, then almost everyone at least noticed the scene and could answer basic, nonspecific questions about it, such as whether the scene was of a beach or a mountain. But if they simultaneously had to perform a very attentionally demanding task, such as keeping track of a set of moving objects superimposed on the images, then only 12 percent of the subjects noticed any scene whatsoever. This clearly shows that we need at least to allocate some attention in order even to get a weak impression of a scene.

And in fact, for any contents of awareness you'd care to name, including tremendously simple features—such as a colored dot, or the angle of a simple patch of grey—if you carefully and fully divert attention away from the feature, it fails to enter consciousness.

So attention is certainly a necessary gating component of consciousness, and while full consciousness of some detail means it has to be strongly attentionally favored as it firmly enters our limited working memory, the same working memory holder can weakly store an approximate group of items that we are faintly conscious of. Occasionally our entire working memory is even called upon to recreate a brief glimpse of a scene, but the lack of detailed analysis or pointed attention is reflected in our very imperfect awareness of the features in front of us.

CHUNKING AND CONSCIOUSNESS

Now returning to chunking, although this process can vastly increase the practical limits of working memory, it is not merely a faithful servant of working memory—instead it is the secret master of this online store, and the main purpose of consciousness.

So far I've argued that attention is the gatekeeper of awareness. Sometimes it chooses what enters consciousness because of pressing biological issues, such as a potential danger, and sometimes it chooses what enters based on a deliberate goal we have set ourselves. But whatever enters consciousness reflects a first guess, a provisional analysis, that this item is currently very relevant to us, based on our various needs. And the output of attention and the arena for consciousness is our working memory, which is limited to a maximum of about four or so items. But, crucially, all those objects are processed as deeply as our brains allow, and this makes every detail of every item available in a unified way. We are then free to apply various strategies to further examine the items, notice similarities and differences between them, combine them, swap them around, and so on. I've repeated the mantra throughout this book that consciousness is concerned with information—specifically, useful, structured information. Chunking is the main catalyst within the bubbling cauldron of working memory where we convert the raw dust of data into molten gold, where basic information from our senses joins the highly refined, hierarchical edifice of meaning that we've been building up from birth.

There are three straightforward sides to chunking processes—the search for chunks, the noticing and memorizing of those chunks, and the use of the chunks we've already built up. The main purpose of consciousness is to search for and discover these structured chunks of information within working memory, so that they can then be used efficiently and automatically, with minimal further input from consciousness.

First the search: Surprisingly, the straightforward result that working memory is limited to four items was only accepted relatively recently, at the turn of the twenty-first century. For the entire half century preceding this, most psychologists assumed that our working memory capacity was around double this, mainly because most researchers failed fully to acknowledge how ubiquitous human strategic processing is and how we all, as a matter of course, use these strategies to boost performance.

I have attended hundreds of research talks over the years, and at these seminars I've heard a particular complaint again and again. It can involve any lab, almost any kind of experiment, and any human population group: However tightly you try to control an experiment, if it poses any kind of challenge to the subjects, those pesky human volunteers will almost always find some strategy to improve performance, usually in a way that neatly makes the experiment invalid. Human innovation is not confined to inventions for revolutionary cyclonic vacuum cleaners and state-of-the-art tablet computers—it is happening almost all the time in all of us, whenever we are awake. Searching for and then finding useful strategies for solving problems, whether large or small, is a signature feature of consciousness.

Perhaps what most distinguishes us humans from the rest of the animal kingdom is our ravenous desire to find structure in the information we pick up in the world. We cannot help actively searching for patterns—any hook in the data that will aid our performance and understanding. We constantly look for regularities in every facet of our lives, and there are few limits to what we can learn and improve on as we make these discoveries. We also develop strategies to further help us—strategies that themselves are forms of patterns that assist us in spotting other patterns, with one example being that amateur track runner developing tactics to link digits with running times in various races.

One problematic corollary of this passion for patterns is that we are the most advanced species in how elaborately and extensively we can get things wrong. We often jump to conclusions—for instance, with astrology or religion. We are so keen to search for patterns, and so satisfied when we've found them, that we do not typically perform sufficient checks on our apparent insights.

And we really are a decidedly strange species for actively seeking out games with patterns in them, when such activities seem to serve no biological function whatsoever, at least not in any direct way. It's as if we were addicted to searching for and spotting structures of information, and if we do not exercise this yearning in our normal daily lives, we then experience a deep pleasure in artificially finding them. Crossword and sudoku puzzles are obvious examples, but there are many other types of games, quizzes, puzzles, and so on that inject pleasure into our lives because of the sense of satisfaction we feel when we spot beautiful structures.

And there are some particularly mad people, such as myself, who even decide to dedicate their careers to searching for patterns within a set of information. Scientists may well be motivated to improve society by their discoveries, but most are largely driven each day by their curiosity—a desire to convert some messy pile of experimental data into an elegant, neat little explanation. When describing what I do to those outside of research, I regularly find the most useful analogy is that science is like trying to solve a huge, fuzzy crossword puzzle.

But hobbies in searching for patterns are not by any means limited to the sciences. The arts, too, generate their richness and some of their aesthetic appeal from patterns. Music is the most obvious sphere where structures are appealing—little phrases that are repeated, raised a key, or reversed can sound utterly beguiling. This musical beauty directly relates to the mathematical relation between notes and the overall logical regularities formed. Some composers, such as Bach, made this connection relatively explicit, at least in certain pieces, which are just as much mathematical and logical puzzles as beautiful musical works.*

But certainly patterns are just as important in the visual arts as in music. Generating interesting connections between disparate subjects is what makes art so fascinating to create and to view, precisely because we are forced to

---

*In Douglas Hofstadter's whimsical and influential book *Gödel, Escher, Bach: An Eternal Golden Braid*, these logical structures in music and art are carefully explored, especially in terms of how they relate to the human mind.

contemplate a new, higher pattern that binds lower ones together. And literature is most powerful when it is using exactly the same trick.

We are alone in the animal kingdom in just how aggressively we constantly search for patterns, and even in how they may be a source of much of our pleasure, both in our creative acts and in our more receptive appreciation of certain hobbies.

The second aspect of chunking is the actual detection of these chunks of structured information within consciousness, and this is where the most central purpose of consciousness resides. What do I mean by patterns, structure, or regularities (words that I use interchangeably)? To take an extreme example, a million 1s and then a million 0s can either be seen as 2 million pieces of information, which would take about three times the length of this book to write out, or we can simply transform those 2 million items into just a single sentence by saying, "It's just a million copies of 1 and then a million copies of 0." In other words, spotting patterns is about finding redundancy in the information. You can *compress* the information into a different, smaller, and more useful form by spotting parts that repeat in some way or other, and, ideally, capturing the repetitions in a rule.

If we can successfully turn any group of data into a pattern or rule, then near-magical results ensue. First, we no longer need to remember that mountain of data—we simply need to recall one simple law. But the benefits don't just stretch to memory. We're also, crucially, able to predict all future instances of this data, and so control our environment more efficiently. The rule may even capture something about the *mechanism* of the data, allowing us to understand it in a more fundamental way.

Say I'm a prehistoric man. Winter is closing in and my family and I are starving. We dig up some potatoes, which at first sight look edible. We take a bite and spit it out in disgust. I chuck the potatoes on the fire, out of rage. A few hours later, when the fire has died down, I pick the potato up, feel that it has softened considerably, and take another bite. It's delicious. A few months later, exactly the same set of events happens. I try a raw potato, chuck it in the fire in rage, then a few hours later try again to find it not only edible but delicious. Now I could just remember these as two completely separate episodes in my life and move on. *Or* I could detect a possible pattern—repeated connected instances between food and fire, with fire making potatoes edible. I then no longer need to remember what the weather was like in each of these instances, where I placed my fire, or what words my children said at the time. All that matters is that fire makes potatoes edible. Now that I've detected a pattern, I can consolidate the crucial elements of my memory, thus lowering the load; apply this memory chunk to all future instances of fire and potatoes; and start thinking about other similar organic objects that might benefit from a similar treatment—perhaps sweet potatoes or squashes. By crystallizing and compressing my memory, I've actually gained a considerable amount of power over the environment, and my family is less likely to starve as a result.

Some of our greatest insights can be gleaned from moving up another level and noticing that certain patterns relate to others, which on first blush may appear entirely unconnected—spotting patterns of patterns, say (which is what analogies essentially are).

It's difficult to overestimate the extent of learning that is captured by chunking processes. The example of the volunteer who heard digit sequences for two years dramatically illustrates how chunking can vastly increase our short-term memory capacity. But chunking is the main process that we consciously use to turn *any* novel pattern into a structured part of our memories, which is what almost all learning involves. Our long-term memory store, with thousands of related items, is usually unconscious, waiting loyally for our conscious minds to retrieve some item. But at one point in the past each item was met anew, consciously labeled as important, as relating to other features of our preexisting knowledge, and laid down in memory. The initial stages of learning are always the hardest, but once the first foundations are built, we can connect new items with what we've already memorized; as the tapestry of knowledge builds, it becomes ever easier to learn a new part of a topic, because it increasingly connects to related memory items. Closely connected individual items form chunks together, which then connect up themselves in ever larger bound objects in memory. In this way, we can use consciousness—and chunking—to create a highly functional, hierarchical, interrelated bank of knowledge, where, by the time we reach adulthood, most seemingly novel items have some preexisting context. And these heavily embedded prior expectations from the fruits of our vast learning can in turn heavily guide our attention to decide what to load into our working memory, furthering our chances to discover in awareness something novel or important by which to incrementally improve our world model.

So, chunking within working memory is both the arbiter and the indexer of our long-term memory store, always striving to make what's most important to us most easily accessible, and to forge new groups out of apparently independent items, based on the patterns we discern. Consciousness and chunking allow us to turn the dull sludge of independent episodes in our lives into a shimmering, dense web, interlinked by all the myriad patterns we spot. It becomes a positive feedback loop, making the detection of new connections even easier, and creates a domain ripe for understanding how things actually work, of reaching that supremely powerful realm of discerning the *mechanism* of things. At the same time, our memory system becomes far more efficient, effective—and intelligent—than it could ever be without such refined methods to extract useful structure from raw data.

LANGUAGE—JUST ONE KIND OF CONSCIOUS CHUNKING?

Language is a special case, since, by learning a language, we can exponentily increase our capacity to learn more generally via the organization of our own thoughts, via books, teachers, the Internet, and any other form of human communication that can transmit well-chunked information. Whether any other species can start to learn some components of a grammatical language is a controversial question. What's not controversial is that even if other animals can learn some aspects of language, humans dramatically outshine other species in this sphere, picking up both a vocabulary of thousands of words as well as a complex grammatical structure. All of this is due to chunking.

As we learn language for the first time, we constantly attempt to make successful inferences between the sound of the word we've just heard and the key feature of whatever else we've just experienced through our senses. We slowly build up our language, with many words initially starting as overly general chunks. For instance, our baby daughter started saying "Mama" quite early on, which initially excited my wife to no end, until we realized it merely meant she was generally unhappy. But language soon allows us to use the linguistic objects we've already learned to hone our skills further, building up our rich web of meaning.

Many psychologists believe in a "language instinct"—a set of uniquely human brain regions and mental skills built especially for language. My view instead is that language emerges out of our more general capacity to make conscious chunks. As consciousness and the ability to chunk begin to flower, this allows the child to extract these language chunks from what she hears, just as she's learning about many other types of structured information, such as how to walk, how to deal with complex social situations, or how to interact with her more advanced toys. At the same time, adults learning an artificial grammar, akin to acquiring the rules of a novel language, do not activate some special language region. Instead, they activate just the same brain areas as when they are performing any other chunking task, such as encoding structured spatial sequences. And one would expect a language instinct to have genes associated with it, giving us a specific helping hand from our earliest days. To date, there is controversial and patchy evidence for this.* Instead, we have a sufficiently advanced form of consciousness that is hungry for patterned, hierarchical information and playfully active in its search for powerful, structured cognitive tools to manage this information—of which language is perhaps the most rich, useful example. After all, language allows yet another level of information processing in nature, where chunks of insight can be passed between people, and collaboration between members in a group can generate innovative new ideas, which would not have been possible alone.

THE FRUITS OF CHUNKING AND AWKWARD SELF-CONSCIOUSNESS

The third and final aspect of chunking in relation to consciousness is what we do with those chunks when we've firmly acquired them. In this case, it's generally better if we're *not* aware of them—at least not anything other than the absolute top-level chunk we're currently dealing with.

The main purpose of awareness is either to manage tasks too new or complex for our simpler unconscious mind or to innovate, to find patterns

---

*The main neuroscientific evidence for the language instinct comes from the discovery of a gene called FOXP2. A mutation in this gene, it has been claimed, causes a selective language impairment, especially in the ability to speak fluently. But the mutation also causes general cognitive deficits, such as a lowering of IQ, which might underlie an impairment in the kind of chunking processes I'm describing rather than anything specific to language.

in our working memory, so that we can optimize and automate biologically relevant goals. But once those tasks have been heavily learned, if they take up consciousness again, then that is far from ideal. First, because we are simultaneously analyzing the details of the task, we are likely to perform it less efficiently than when it was an automatic habit. Then there is the issue of energy use: Consciousness requires a large and active network of cortical regions, which in turn need considerably more energy than unconscious, automatic habits. So it is wasteful to use consciousness for anything other than its official purpose, to discover potentially significant opportunities to improve our mental programming.

Some chunks are too complex for our unconscious minds to handle and instead serve to guide our consciousness toward a specific goal. For instance, if I want to make a relatively nontrivial change to my PC, such as modifying the monitor resolution, I know exactly what sequence of mouse clicks to make to achieve this, and each one has to be consciously controlled. But the vast majority of these well-learned packets of information can whirr away quite happily while conscious engagement is entirely elsewhere—for instance, me walking for many minutes, oblivious to my surroundings, or driving while daydreaming.

It's very much as if we had two modes: a slow, deliberative, highly conscious system, there to detect novel or ever more complex forms of patterned information, to find its structural essence, which we use to build chunks; and a fast, automatic, barely conscious system, which takes advantage of the well-honed chunks that consciousness has previously formed.

This dual mode is highly reminiscent of the balance in nature between well-established genetic traits and chaotic innovation. Recall that if an organism is coping well, then mutation rates are lower. During these times of stability, creatures such as the nematode worm will rely on the more faithful replication mode of cloning. They are in effect banking on their previously effective DNA-based beliefs about the world, and resisting attempts to change. But if there are severe stresses to life, and innovation is required to ensure survival of offspring, then random mutation rates or sexual selection will become more likely, as if the species were attempting to breed a novel successful genetic trait in the next generation. In other words, a more dynamic, daring approach is needed in order to find a tangential solution to the current problem.

The similarities between these genetic processes and consciousness are more than skin deep. They relate to the general question of where you sit on an information-processing continuum between stability and chaos. It's a good policy to err on the side of stability if things are going well, and bank on those previous insights that are currently working, but to verge toward the chaotic, innovative side when life is dangerously threatened.

The interplay between conscious chunk formation and unconscious automatic processes is an attempt at simultaneously exploiting both sides of the information-processing coin. One way of viewing consciousness is as an innovation machine, there to dabble with the chaotic side when at an impasse and new ideas are required to improve matters. But, unlike the genetic tricks, consciousness carries out these semi-chaotic probings in a relatively safe and highly directed way, only accepting those new insights that will clearly benefit the animal, either in terms of adding to its world picture or enhancing its behavior. Then, once these innovative chunks have been discovered, they are incrementally added to the stable portion of cognition, which largely resides in the unconscious.

This view highlights the evolutionary advantage of awareness. Engaging consciousness admittedly eats up resources, as the most energy-hungry parts of the brain are engaged. However, the energy savings that ensue far outweigh the initial costs, since existing tasks become efficiently streamlined and new, intelligent techniques are discovered that allow us to avoid complex threats and obtain challenging rewards.

While this division of labor is fantastically effective on the whole, we can really screw things up by mixing these two modes. This happens when we deliberately try to be conscious of well-worn chunks of skill or memory. It's important that we can open up our consciousness like this on occasion, at least partially. This way we notice faults in deep habits and can improve, even if we are likely to interfere with performance while the reassessment is being made. But many times, this mixture just produces, at best, semi-defective results. It is a mode we colloquially label as being "self-conscious."

For instance, if you start attending to every little movement of your tennis forehand stroke, the fluidity falls apart. You are overloading your working memory far beyond its capacity and it becomes overwhelmed. To take an exaggerated view, you might place in working memory just the first, eighth, twelfth, and fifteenth component of the muscle sequences, and this will result in a very awkward motion, hardly resembling the fluid forehand you demonstrated without much awareness a minute ago.

You have also reentered a state in which you are questioning and analyzing every aspect of your movements, a state that is ravenous for new information, for novel, efficient patterns of thought and behavior. You are temporarily rejecting the information you've built up about your structured movements, and instead are searching for patterns anew. You would be far more aware of any movements now and might be able to reconstruct or tweak your forehand technique, but at the expense of having the previous elegant chunk of forehand motor memory at your ready disposal.

Ample evidence from the lab supports this dual-mode view, where consciousness is initially necessary for complex learning, but then largely gets in the way of automatic processing once the lesson has been firmly acquired. For instance, in a series of elegant experiments, Sian Beilock and colleagues tested how golf-putting performance can be manipulated according to where we point our attention. If expert golfers are told to focus on the swing of their club, then the ball ends up further from the hole compared with when their attention is distracted by another task, namely, listening for certain target sounds amid a stream of beeps. This set of results is the exact opposite of what happens with novices, who are far more accurate in their golf swings when focusing on the movement of the club, compared with trying to listen for these target beeps. Very similar results have been found in soccer, baseball, and even touch typing.

But leaving these uncommon glitches of our conscious system aside, the fruits of chunking do allow us to excel at almost every field, as long as we put in the time. Many hours of practice, directed in the right way, enables our working memory to spot increasingly subtle, sophisticated patterns relevant to the task.

Our ability to consciously apply our chunking skills, both by detecting crucial regularities in a sudden burst of insight and by more patiently, steadily building up layers of structured knowledge over months or years, is essentially responsible for every human advance and every intellectual achievement in our history.

Our collective curiosity, working intimately with our prodigious talent for noticing patterns in our world, has yielded many incredible scientific insights and technological marvels. And, I would argue, our most cherished artists, writers, and musicians couldn't have produced their gifts to culture without their skills for seeing the hidden structures around them. Geniuses in these fields should perhaps be defined not only by raw ability, but also by their years of painstaking, focused, conscious attention, which allows them to detect and construct deeper chunks than the rest of us.

The Brain's Experience of a Rose

Neuroscience of Awareness

QUIT WHILE YOU'RE AHEAD?

Recently, I was fortunate to attend a two-day symposium on consciousness and cognition organized by the Royal Society, the UK's main scientific society. A select group of leading lights from neuroscience around the world descended on a small village in the middle of England. They came to give various talks on the progress science is making in unlocking how the brain creates consciousness. Their audience included members of the society as well as people working closely in the field, like me. The location was Chicheley Hall, a beautiful English mansion owned by the Royal Society that it uses when hosting mini-conferences like this one. With exquisite gardens around us, decadent food, and plush accommodation, the atmosphere was surprisingly informal.

One of the speakers, Michael Gazzaniga, is one of the old grandees of cognitive neuroscience. He's been highly influential for many decades in the field of neuroscience, not only in terms of research and teaching but also in communicating its discoveries to a wider audience. Now in his early seventies, he's a tall, charismatic, friendly man with an easy, welcoming laugh. He began his own talk by relating what had happened to him a few days earlier on entering the country. Arriving at London's Heathrow Airport, he made his way to passport control. The passport officer asked him the standard question: whether he was there on business or pleasure . . . business was the answer, for a few days, to give a talk. The officer asked him what business he was in. He explained that he was a brain scientist, which caused the officer to raise an eyebrow. This officer, increasingly intrigued, said, "You mean like the right brain does spatial stuff and the left brain does language?" Feeling both lucky and rather proud, Gazzaniga replied that, actually, he'd had a part to play in establishing this result many years ago. The officer, impressed, asked Gazzaniga what he was going to talk about this time. Gazzaniga replied, "Perspectives on consciousness and the brain." The passport officer gave him a sideways look and a frown before suggesting, "Have you ever thought of quitting while you're ahead?"

It isn't just the public that cocks a suspicious gaze toward the neuroscience of consciousness. Many neuroscientists feel the same way themselves, even now. And yet, as I will relate in this chapter, much exciting progress has been made in the field in the past two decades, and researchers are beginning to converge on a view of what brain areas are involved in consciousness, how these regions interact to generate our experiences, and the signature neural features that mechanistically explain awareness.

My Conscious Mind Is My Conscious Brain

In Chapter 1 I recounted how my friend Martin Monti had employed the fMRI scanner, with me inside it, as possibly one of the most expensive, most cumbersome chat tools on earth. And yet, it demonstrated in a relatively elegant way that my conscious thoughts were physical thoughts, that my brain was the specific location where my consciousness resided. Any philosophical position that opposes this stance is balancing on quicksand.

Not only do we manipulate our consciousness every day by interfering with brain function, by way of the drugs we consume, such as caffeine or alcohol, but countless medical examples make the link even more obvious. For instance, many different forms of brain damage cause profound and lasting changes to consciousness.

If we assume that consciousness is simply a physical process generated by brain activity, then this makes the mystery of awareness all the more vivid and exciting; it gives us a provisional scaffold to examine so that we can try to find answers. Linking consciousness with the brain admittedly raises a new set of questions, but now they are pointed and open to scientific inves-

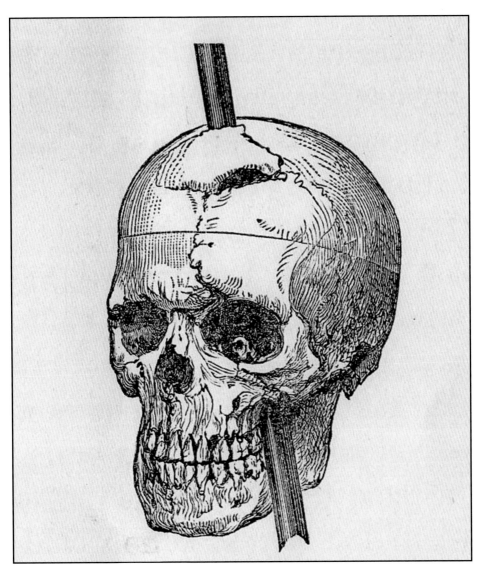

Figure 1. Harlow's illustration of the path that the tamping iron took through Phineas Gage's skull and brain.

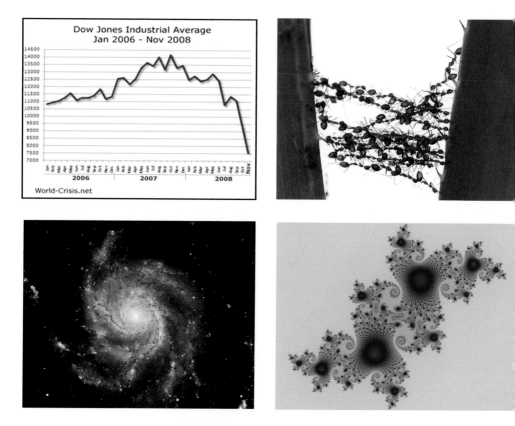

Figure 2. Various examples of complexity arising from simpler underlying structures, rules, or behaviors. *Top left*: the collapse of the Dow Jones Industrial Average following the 2007–2008 credit crunch. *Top right*: ants working together to form a bridge across a gap that would be too great for any individual ant to traverse on its own. *Bottom left*: an example of a spiral galaxy. *Bottom right*: a Julia set fractal.

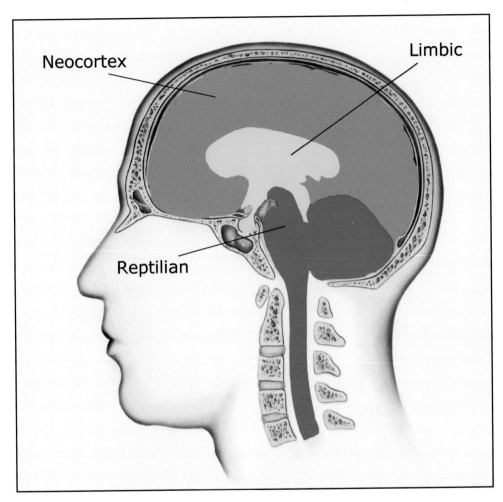

Figure 3. Schematic of the human brain, with the most primitive ("reptilian") brain in the center, the limbic ("early mammalian") system surrounding this, and the ("late mammalian") neocortex making up the outer shell.

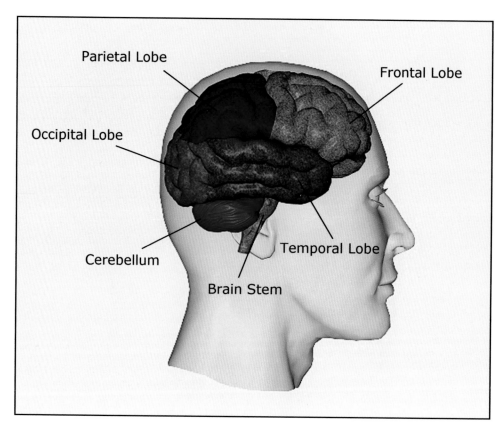

Figure 4. The four lobes of the human brain.

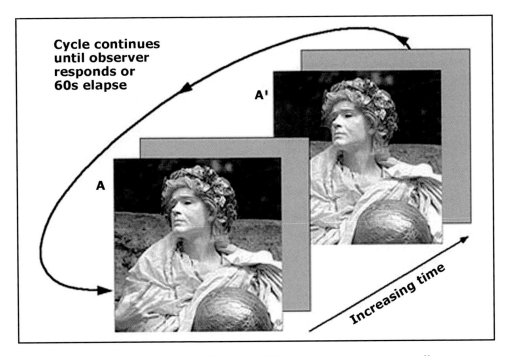

Figure 5. An example of change blindness. The two figures are repeatedly swapped, in between a blank grey screen, until the volunteer spots the blatant but unexpected change.

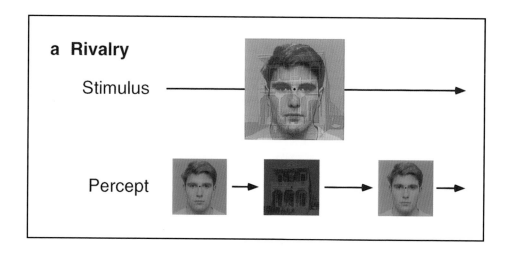

Figure 6. Examples of stimuli that induce repeated switches in visual perception. In the top example, involving binocular rivalry, a single mixed picture presents a face to one eye and a house to the other, because of red/green filtered glasses. The experience flips between only a face and only a house randomly over time. In the bottom two examples, we experience either only a candlestick or only two profile faces (left), and either only an old woman or only a young woman (right), and our experience again flips back and forth randomly over time.

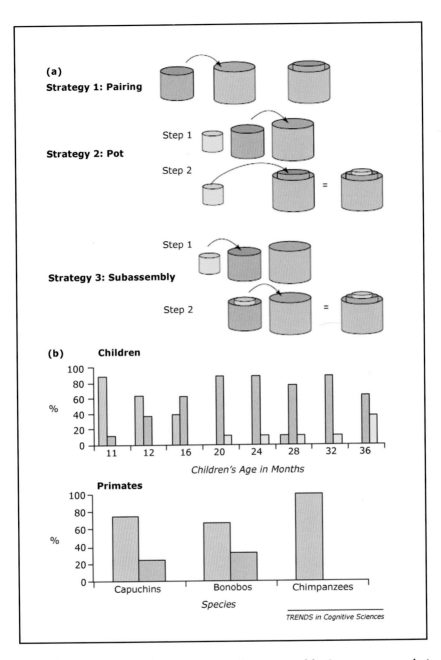

Figure 7. (a) Illustration of the three types of cup assembly. Strategy 1 merely involves putting one cup inside another, and not completing the puzzle. Strategy 2 involves putting the cups inside each other one at a time, and never moving a cup with another inside it. Although the puzzle can be completed in this way, it's different from the demonstration that the experimenter showed the participant. Finally, there is strategy 3, which involves some level of hierarchy, as both the smaller and middle cups are moved simultaneously to fit into the larger cup. (b) Graphs to show dominant strategy used for infants and animals—strategy 1 in blue, 2 in magenta, and 3 in yellow.

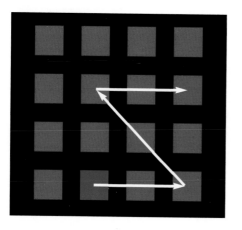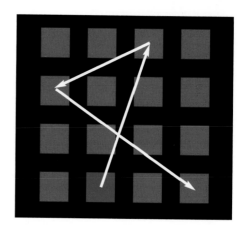

Figure 8. In the fMRI scanner, normal volunteers see a 4 by 4 grid of red squares. Four of them blink blue in a sequence, which the volunteers have to remember for a few seconds. The left side is an example of a structured sequence, and the right side is an unstructured example.

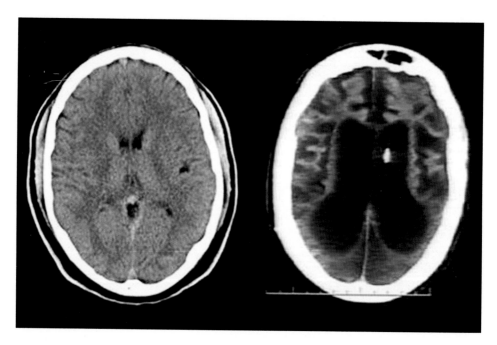

Figure 9. A CT scan comparison of a normal brain, on the left, and Terri Schiavo's brain, on the right.

tigation: Does the whole human brain generate consciousness—or a few specific regions? If only a subset, what differentiates these from the others? Is consciousness related to some particular way that neurons can talk to each other, or is this irrelevant? Does the number of neurons matter, or can consciousness reside in just a handful of nerve cells? Are there different kinds of consciousness for different brain regions, each processing distinct functions, such as our perception of faces and language, or is there only one form of consciousness? And does consciousness have to be tied to brains, or can it be supported by other suitable physical structures—for instance, a silicon-based computer? This chapter is devoted to answering these questions. But it will also continue the work of the previous chapter by making broader claims about the purpose, composition, and mechanism of consciousness.

The study of how the brain creates consciousness involves two surprisingly disconnected wings. The first side has explicitly sought to discover the brain areas and neural mechanisms responsible for consciousness. But in parallel to this strand of research, a largely separate group of scientists has been revealing equally insightful discoveries about the neural recipes for our experiences, despite the fact that they rarely even mention consciousness. Instead, their topics of investigation include the neuroscience of working memory, attention, and chunking.

In this chapter, I will recount the direct, official evidence for how the brain generates our sense of awareness. But then I will show how the study of the neural underpinnings of attention, working memory, chunking, and related topics map closely onto the "official" set of consciousness data, providing further evidence for the importance of these processes to consciousness. In the final section of the chapter I will describe current neural theories of consciousness to provide deeper insights into its nature, especially when strengthened by a synthesis of the two experimental stories.

OPENING THE FLOODGATES

Probably the simplest question to ask of the neuroscience of consciousness, and therefore a natural first point of attack, is whether the whole brain contributes to consciousness, or merely a few of its key areas. There is an unfortunate abundance of data on this question in the form of patients with damage to different brain regions. En masse, this population of patients has localized damage to every brain region there is. This allows us to know, for instance, that the cerebellum, part of the ancient reptilian section of the human brain, has little to do with awareness. Patients missing a cerebellum show no clear impairments of consciousness. For instance, one woman was born with almost no cerebellum in either hemisphere, but she was able to lead a relatively normal life, holding down a job at an electronics factory.

So certainly not all brain areas equally contribute to consciousness. The above example also partially answers another question from the list—about consciousness simply relating to number of neurons. If neuron count equated to levels of awareness, then the cerebellum, with perhaps 80 percent of the entire brain's neurons, would be the most conscious part. That clearly isn't the case.

In fact, the most critical region for consciousness is also one of the smallest and—in some ways—the least interesting, for the purposes of this book. This brain area is known as the reticular formation. It is part of the brain stem, the most primitive of brain regions and another component of the reptilian brain. The reticular formation controls the sleep-wake cycle through a complex set of subregions that each play a part in a chemical and neuronal cascade of activity. These actions allow us both to wake up and to enter different modes of sleep. For instance, when we dream, it is the reticular formation that pushes signals down to the spinal cord, actively paralyzing the rest of the body so that we don't start actually crashing into walls when we're dreaming we're running around. Damage to the reticular formation is usually a pretty absolute business. Either the patient will die or he will be totally robbed of consciousness in a deep coma, unable to ever wake up.

But while we definitely need our reticular formation in order to be aware, it doesn't follow that our consciousness actually occurs in this primitive brain area. My PC will simply not turn on without the power supply unit, but this box is the dumbest part of my computer, having nothing to do with the processing that gives me a functioning operating system.

A somewhat more relevant brain region is one of the main output regions for the reticular formation, the thalamus. It sits just above the brain stem, right in the center of the brain. This particular region is very special because it acts as a hub, the Grand Central Station of the brain. Its neurons receive inputs from, and send outputs to, almost every other brain region. Historically, it was viewed mainly as a sensory conduit, because the thalamus is usually the first port of call for the senses before the information is shunted to the cortex. For instance, for vision, our eyes report what they see via a thick information highway known as the optic tract, which flows into a part of the thalamus known as the lateral geniculate nucleus. This then shunts the information on to the primary visual cortex. More recently, though, the role of the thalamus has been considerably upgraded from passive relay. It is now known to be a sophisticated information-filtering and -organizing device and is thought to play a central role in consciousness.

Patients with extensive damage to the thalamus tend to enter a so-called vegetative state. This is distinguished from coma in that vegetative patients show signs of wakefulness—they may open their eyes more in the daytime, for instance. It's as if their reticular formation is still partially carrying out its duty of waking up the brain, but there isn't enough neural coherence for them actually to be aware of anything when they are awake. As the thalamus is a key player in generating consciousness, it's natural that its functions have been discussed both in terms of information processing and attention: Along with helping with the flow of information between most regions of the brain, the thalamus helps to point our attention in various directions. This explains how those in a vegetative state with damage to their thalamus can appear to be awake, but lack any directed consciousness. Imagine a life where you were alert in some sense, but your attention was never directed toward any object, any goal, any thought—would we even call such a state conscious?

Returning to the e-mail analogy, the thalamus in our large company would take on the role of the IT department. All e-mail coming in from outside (sensory input) first filters through the servers in this large department before being sent on its way to the other departments for which it was intended. And most internal e-mails also pass through the buzzing IT servers in this center of the building before finding their correct recipients. But this IT department is a particularly proactive one. It tries to clamp down on spam before it even turns up in anyone's inbox. And if the situation calls for it, the IT managers won't hesitate to delay any e-mails relating to a particular subject—say pop-music listening, while adding "PRIORITY" to the subject headers of any e-mails that relate closely to the main company project of the moment.

If the IT department servers crash (the thalamus is damaged, causing a vegetative state), there may still be power to the building from the generators in the basement—the lights will still go on in the evening and off in the day-time, and people can still use their computers, and chat in person with their office neighbors to pass on any interesting piece of local gossip. But com-munication won't flow effectively through the building, the large group of managers who make the decisions will largely be ignorant of what is hap-pening, and little real work will get done.

BLINDSIGHT PATIENTS LEADING YOU UP A BLIND ALLEY

It would be incredible, though, if the thalamus were the sole region in the brain that created consciousness. It is a relatively primitive region, present in any animal that has a backbone. Although it is clear that the thalamus is important for consciousness, and is directly involved in producing our ex-periences, it is likely that other regions are also critical.

When the experimental study of the neuroscience of consciousness was in its infancy thirty years ago, there was really only one topic in town: blind-sight. When I was an undergraduate in the mid-1990s, and we examined the question of how the brain supports consciousness, we studied little else. Blindsight still figured prominently when it was my turn to teach under-graduates some years later. Therefore, it would be remiss of me not to men-tion this subject, even if I don't think it casts that much light on the field.

Blindsight is a paradoxical, puzzling neurological condition. If a patient sustains damage to her primary visual cortex, at the very back of the brain, then she will be blind, at least in some part of her visual field. It turns out, though, that in this context "blind" can be an ambiguous term.

The traditional story is that these patients will have no conscious expe-rience in their damaged field of vision. If you move an object left or right in this patient's "blind" region and ask her to tell you which direction it went, she will laugh at you, saying she sees nothing and so obviously can't tell you—for all she knows you didn't even move any object. If the experimenter then asks the blindsight patient to humor him—just to take a random guess, either left or right, even if she doesn't really see a thing—lo and behold, even though she has no visual experience of this movement, she will guess cor-rectly, much to her surprise. So the information is still there somewhere in the brain, but the experience of it, the actual seeing part, is missing. Similar stories occur for "blindtouch," where information about touch sensations is available, even when the feel of them isn't, and so on. Blindsight and similar conditions, the surface story argues, demonstrates a sharp fracture between sensory data and the experience of that data—a potential problem for anyone who links consciousness with information. This traditional story also implies that our experiences are generated from the primary sensory regions of our brain, since losing these regions abolishes our experiences of those specific stimuli.

This, as I say, is a traditional view. The trouble is that it is far too simplistic. First off, most of these patients usually are only partially blinded. Some of the time they do actually have a visual experience and some of the time they don't, even if it's clear that their entire primary visual cortex is missing. Some patients either seem to recover slowly, or get better with practice, generating further doubt as to exactly what level of visual awareness they are reporting. And the sensory information itself isn't perfectly intact—far from it. The patients guess better than chance, but for some patients and some aspects of vision, not by much. In line with this, some patients have been shown, when carefully tested, not to have a total absence of awareness, but merely a degraded form of consciousness, in a way that closely matches their reduced ability to detect visual objects.

So we have a rather more murky reality than the traditional story typically reveals, where there is an indeterminate, shifting level of sensory experience, and a far reduced amount of information available relating to that sense.

Thus there is no strong dissociation between information and experience here. Instead, both are reduced, but in idiosyncratic, fickle ways. It's as if there were faulty wiring in a house, causing only one bulb out of a dozen to turn on in a room—but this single illumination only flickers on and off, creating a dim, indeterminate, shadow-laden view. In this light-impoverished place, sometimes the best that can be achieved is a hunch based on a slightly desperate internal analysis of the meager perceptual input. This may be right a little more than wrong, but you so wish the lights worked properly, and you could clearly see everything around you.

One final point about blindsight is that although it seems at the very least to follow from such cases that the primary visual cortex is the seat for visual awareness, since this kind of experience is still heavily disrupted in these patients, even this conclusion is overreaching. An equally plausible interpretation is that the primary visual cortex is merely an early conduit for the kind of information *that only later in the brain* will become conscious. In analogous fashion, if a person went blind because of damage to his eyes, it would be shockingly premature suddenly to claim that all aspects of visual awareness occurred in his eyes.

The take-home message, therefore, is that although blindsight is a strange condition, it in itself doesn't explain too much about consciousness. But if you view blindsight instead as a patient group that sometimes can be conscious of, and sometimes be blind to, the same kind of visual stimulus, that in itself makes them potentially an invaluable set of people by which to study consciousness. The trick is to control performance, by carefully tweaking the stimuli, so that the patients are usually correct in guessing its direction of motion regardless of whether they are aware of it or not. Then, with everything controlled for, you can place the patient in the scanner with these stimuli and discover which additional brain areas will light up when the patients are conscious of the moving dot, compared with when they can't see it (but can still correctly guess its direction). Larry Weiskrantz, who coined the term "blindsight" in the mid-1970s, carried out exactly this experiment in 1997, along with colleagues, including Arash Sahraie, on one blindsight patient in the fMRI scanner. The main additional activation for the aware condition was the lateral prefrontal cortex, which has already cropped up a few times in this book; its function is closely tied to working memory and other complex cognitive processes. But for the moment, it's worth emphasizing that the patient's main additional brain activity when he was aware of these moving visual objects was in generalist nonsensory regions, those that adapt their function to the task at hand.

VISUAL HIGHWAYS TOWARD CONSCIOUSNESS

Blindsight, although not necessarily informative on its own for how the brain represents consciousness, was useful in opening the floodgates for other researchers from neighboring fields to join the party. In the mid-1990s, monkey vision researchers also became interested in the question of whether the primary visual cortex was really the seat of visual awareness. Blazing a trail in this field was Nikos Logothetis and his team, who carried out a series of groundbreaking studies using electrodes implanted into various places in the monkey brain to record the activity from individual neurons. The monkeys were performing a now classic task in the consciousness literature known as the binocular rivalry test. If you present one image to one eye—say, a face—and a completely different image to the other eye—for instance, a house—then a person does not actually see some chimeric amalgam of a face and house. Instead, the subject will experience only a house, followed at some later time by just a face, and then a little later the visual image will flip back to a house, and so on (see Figure 6, top).

If you give the binocular rivalry test to humans, you merely need to spend a few minutes explaining to the participant that she should press a button to indicate when her experience switches from a face to a house or if the reverse switch occurs. Monkeys can do the same task, except you need months of training to get them to carry it out.

Logothetis discovered that the primary visual cortex wasn't nearly as vital for consciousness as the blindsight evidence initially hinted at. Although primary visual cortex neurons in these monkeys showed an accurate representation of what was actually presented to the eyes, neuronal firing was poorly related to what the monkey "perceived": Only a fifth of the neurons seemed to match the contents of the monkey's reported "experiences." So the primary visual cortex, rather than the seat of visual awareness, is instead mainly a filtered copy of what comes into the eyes. We may lose much of our visual consciousness if we lose this brain area, but that's merely because the primary visual cortex is the first and main cortical stopover for visual information, regardless of whether or not it is conscious.

To reinforce this point: Have you ever wondered why the world doesn't go black every time you blink your eyes? Whenever you blink, activity in the primary visual cortex massively reduces to reflect the darkness. However, more advanced visual regions, whose main purpose is not to reflect the visual world, but to explain and predict it, can generate a sense of visual continuity by allowing our most recent perception of our surroundings to persist during the momentary gap of the eye blink.

Instead of its role in awareness, the primary visual cortex performs basic processing to pick out the crude raw features of the visual world. It carries a map of that world in an organized way, relating closely to what the eye sees.*

It's worth pausing for a moment for a quick primer on the human visual brain. Let me return to the manager responsible for yellow in the large company, since he's very much in the middle of all the action and has a particularly good perspective on the visual system:

Once the company receives information from the security cameras at the front of the building, above the main entrance, large, thick cables take the data from these two cameras (eyes), via the IT servers in the middle of the building (thalamus), to the back of the building, where the grunts sift through it (primary visual cortex). These are not the most advanced employees in the world, which is why they're paid minimum wage, and they never give orders—they just carry out basic tasks and accept simple instructions. Each grunt is responsible for a single pixel coming from the cameras. The grunts even sit in a vertical scaffold, so that the top right grunt works on the bottom left pixel, the one below him works on the camera pixel above that and so on. If they all held up a color based on what they were looking at, you could almost see a picture from the overall scaffold.

These grunts send a flurry of e-mails to slightly better paid workers who are still toward the back of the building, but in offices in front of them. I'm one of these people. I don't care about single pixels—that's totally beneath me. I manage about a hundred pixels. So a hundred grunts, each responsible for a single pixel, constantly send me e-mails on whatever colors the cameras have seen in that location. And if yellow comes up from enough of the grunts, I get really excited and send out my e-mails. Offices near me get info about movement in just the same way, and they also don't care about single pixels anymore, but a small region of space.

We send out our e-mails to offices further to the front of the building. As I said before, if I send e-mails out about yellow, and neighbors send e-mails

---

*Actually, the primary visual cortex is not quite as dumb as all that. It can also act as a flexible slave system to more advanced regions of the brain. For instance, attention can enhance our perception of one part of space, partly via later regions controlling the activity of the primary visual cortex, so that those subregions within it, say, that code for the upper right quadrant, are more active, and so more ready to pick up changes in this location.

about black, then there's some guy we e-mail who gets excited about this potential striped spy nearby. So he actually doesn't care where in space something happens—it could be absolutely anywhere in the picture—he just cares about what it is.

So the primary visual cortex right at the back has neurons that each code for a tiny region of space, and collectively you could recreate what the eye sees by a closely corresponding map of primary visual neurons. These primary visual neurons then send out their information to slightly more advanced, later visual regions, a little more to the front, which will be combining this cruder data together to start extracting the important features. A general rule is that the further forward you go in the visual system, moving away from the primary visual cortex, a given neuron will represent more refined information about an ever larger region of space, as increasing levels of data grouping and meaning extraction occur. Here is the factory floor of visual chunk recognition.

From the primary visual cortex right at the back, part of the information might go forward to V4, which processes color, with each neuron coding for a slightly larger region of space than in the primary visual cortex, and then a few more steps will lead to one of the endpoints at the inferotemporal cortex, much further forward in the brain, already in a different lobe, the temporal lobes, where some forms of object recognition occur. A neuron may activate now if a chair, say, turns up in any region of the left half of space.

How does a neuron in the inferotemporal cortex represent the notion of a wasp? Aside from the fact that thousands of inferotemporal neurons will collectively be involved in this memory, these advanced neurons probably store their information via a hierarchy of connections with earlier visual regions. For the inferotemporal cortex to represent a wasp, perhaps it needs to have neurons connected with those for yellow and black in V4, those for a furry texture in V3, and so on.

Logothetis sampled some of these later regions in his monkeys performing his binocular rivalry experiment and found some surprising results. A step or two removed from the primary visual cortex is V4, which processes color, and MT, which processes moving images. Both of these areas were far more reflective of what the animal actually perceived, with nearly half the neurons here involved in the specifics of consciousness. But these were trumped by the inferotemporal cortex, which has nearly all its neurons flipping in activity according to whether the monkey is seeing one object or another.

SCANNING CONSCIOUSNESS AS CANDLES BECOME FACES

Therefore, the more specialized and advanced the visual processing region, the more refined and compound is the information it represents, and the greater is its role in consciousness. But the trouble with these single neuron electrode studies is that they are only sampling a tiny subset of the brain. So you can't make any other claims about any other areas—for instance, the lateral prefrontal cortex mentioned in the blindsight imaging study above—as you have no data on them either way. Only brain-scanning, although far less focused than single neuron electrode studies, can look at the whole brain simultaneously and investigate which suite of regions are involved in consciousness.

In humans, fMRI studies have again shown that the later, more refined visual regions, as opposed to the primary visual cortex, activate when we experience switching visual views. But now two other regions also light up at least as brightly: the lateral prefrontal cortex and the posterior parietal cortex.

Binocular rivalry is but one of a range of visual experiments where our experience flips back and forth between two competing visual views (see Figure 6, bottom, for examples). For instance, there is the famous image of either a candlestick, or two faces in profile, where we never see both the candlestick and faces, only one or the other. Just as in the binocular rivalry experiments, if you record when people's experiences switch, in this case to the candlestick or faces, you again see activity in a combination of later, advanced visual regions, and the prefrontal and parietal cortices—two areas that have no devoted interest in vision at all.

Although little has been said before now of the posterior parietal cortex, it is almost always activated along with the lateral prefrontal cortex, forms a tight network with it, and almost certainly carries out a similar functional role. In fact, these regions coactivate so frequently that some people have assumed they are just one large processing network for advanced, flexible thought. Although there may be subtle functional differences between the posterior parietal cortex and the lateral prefrontal cortex, for the rest of this chapter I'll assume that they are a unified "prefrontal parietal network."

So far, I've only looked at fMRI studies where the test subject's experience switches between one view and another. Many studies have also explored how brain activity changes when we detect a stimulus compared with when it's invisible. For instance, Stanislas Dehaene and colleagues showed subjects a rapid sequence of jumbled squares with nothing of note in them. In the middle of this sequence, though, there was a word. Sometimes the gap between the word and the next set of jumbled squares was too short for the volunteers to notice the word, and sometimes the gap was just long enough for the volunteers to be aware of the word. In the aware condition, compared with the unconscious one, the standard activation pattern occurred, with advanced sensory regions lighting up, along with the prefrontal parietal network.

These results are not limited to vision. You see exactly the same combination of activity in advanced sensory regions and the prefrontal parietal network when subjects detect a touch or a sound, or even when they must use a combination of senses.

It's always useful in establishing a result to make your attack from as many different roads as possible and show that each one leads to the same destination. So another technique increasingly employed is not to ask about how activity in a particular brain region is linked with a given function, but to examine the physical size of that brain structure and see if that relates to some behavioral measure. Recent studies have done this with consciousness, again finding links with awareness and the prefrontal parietal network. For instance, Ryota Kanai and colleagues used another kind of image that flips between two experiences, this time with moving dots appearing to rotate in one direction or another. Kanai found that those individuals who reported more frequent switches in apparent rotation also had thicker parietal cortices—as if having more of this brain makes you notice more changes in your awareness.

Another route of attack is to look at neurological patients: If the prefrontal parietal network is generally involved in all forms of awareness, then damage to these regions should cause some detectable drop in consciousness. That's exactly what researchers are discovering. For instance, Antoine Del Cul and colleagues gave patients with damage to the prefrontal cortex a challenging task where they had to spot a very briefly presented number, which was immediately followed by a collection of letters designed to interfere with the detection of the number. Patients had a reduced experience of these hard-to-spot numbers, even though they weren't much worse than controls at actually identifying the numbers if forced to guess.

You can obtain similar results with the parietal cortex. For instance, Jon Simons and colleagues gave a memory test to patients with damage to their posterior parietal cortex in both hemispheres. First the patients heard seventy-two trivia sentences, such as "Al Capone's business card said he was a used furniture dealer." They had to guess whether the voice they'd just heard was male or female and whether the person behind the voice actually believed the sentence he or she had read out. This was basically to throw the participants off the scent, because, after all these sentences were read out, there was a surprise memory test, where the volunteers had to say whether they'd just heard a set of sentences (half of which were new). These patients were no worse than controls at guessing whether the sentences were new or old, but were significantly less confident in their judgments—as if their awareness of their own memories had faded, even if the memories themselves hadn't.

An intriguing alternative approach is to explore what happens as you slowly sap consciousness from a person—for instance, by varying the levels of general anesthesia. In one such fMRI study, Matt Davis and colleagues played volunteers various sentences in headphones. While the temporal region responsible for the simple and more processed sound components of speech was active regardless of the level of anesthesia, the prefrontal cortex switched off as soon as the volunteers entered sedation.

Outside of the world of imaging, there are fascinating corroborative clues from evolution and the study of comparative anatomy. When researchers measured how much of the entire cortex was taken up by just the primary visual cortex in different primates, they found that humans had the lowest share compared to our primate cousins. Moreover, the acuity of our vision as a primate is poor. Our other senses (and the size of our primary sensory regions) are nothing to write home about either. Our sense of smell is particularly poor, for instance. But evolution has taught us the vital lesson that it's not what you've got, but what you do with it. We take the relatively feeble raw data that comes through our senses and we analyze it brilliantly, deeply, extracting insights continuously. In stark contrast to our shrinking sensory areas, our prefrontal cortex has greatly expanded compared to that of chimpanzees and other primates, almost certainly so that we can extract so much more understanding in exchange for less input. And with the rich discovery of knowledge, so comes consciousness: Whether or not other animals are conscious, it is beyond doubt that human consciousness is the richest, to go along with our highly expanded analytical prefrontal cortex, and intriguingly, despite our diminished sensory regions.

So the general picture here is that although our later, more refined sensory regions are involved in the specifics of our experiences, these regions need to be combined with our most general brain areas in the prefrontal parietal network if we're actually going to be conscious. But at present, while many consciousness researchers have been ingenious at linking different brain regions with awareness, most have stopped there, balking at giving some functional, mechanistic explanation for how these brain areas actually contribute to consciousness. It certainly needn't be this way, and one simple approach is to look to the burgeoning research literature from those who, outside of consciousness research, have sought directly to understand what the prefrontal parietal network contributes to cognition. As I will show in the next sections, adding these clues to the mix is a powerful way to further elucidate the shape of consciousness.

PATIENTS AND THE PREFRONTAL PARIETAL NETWORK'S OFFICIAL JOB

The first case study to link the prefrontal cortex with complex human cognition was famously written up in 1935 by the prominent U.S. and Canadian neurosurgeon Wilder Penfield along with his colleague Joseph Evans. As well as being an important addition to the neuroscience literature, this article carried a moving emotional angle. Wilder Penfield's only sister, Ruth, had suffered from headaches and seizures for many years that were caused by an underlying brain tumor. In order to try to save her life, Penfield saw no other option than to cut out most of her right frontal lobe, and so he performed the surgery himself. The operation initially was a qualified success and bought her an extra couple of years of relatively normal life, but then the tumor aggressively returned. There was little anyone could do. After her death, Penfield felt that she would have wanted to use her experiences to help mankind—and his eloquent description of the changes she went through after losing her right prefrontal cortex certainly did that, as it steered frontal lobe research into a very productive new direction.

Although in many ways Penfield's sister Ruth didn't change following the surgery, she did experience some subtle problems. Her main complaints were that she felt "a little slow" and she could not "think well enough." One specific example in the paper illustrated her problems vividly. She was due to cook a dinner for a guest—in fact, her brother Penfield himself. But although before the operation she would have had no trouble organizing and serving a complex set of dishes, now she found herself flummoxed. She was able to get started on one or two dishes, but after a long and frustrating attempt, she admitted defeat in putting the meal together as a whole. She just didn't know how to organize herself to decide what to chop or heat next.

This situation of being overwhelmed by too many options is highly reminiscent of how things fall apart when we become self-conscious of a previously overlearned skill, such as a tennis stroke. We quickly run out of space in our working memory with all these novel, non-chunked motor commands to coordinate, and our movements become clumsy. Here, likewise, Ruth's main problem might well have been a shrunken working memory, so that even those normal situations and well-chunked sequences no longer fit. The result is sporadic chaos—a failure to perform anything taxing, novel, or complex.

While this report is largely anecdotal, there is good evidence that patients with prefrontal cortex damage do indeed have a working memory deficit. For instance, Cambridge colleagues and I tested a group of patients, each of whom had a lesion in their prefrontal cortex, in one hemisphere on a standard test of working memory. Eight red boxes would appear on the computer monitor in front of the patients, then some of these boxes would blink blue in a sequence. Immediately after this, the volunteers had to touch the boxes in the order that they'd just blinked blue. Compared to a closely matched group with no brain damage, the patients were indeed impaired on this spatial working memory task, especially if they had damage to the lateral part of the prefrontal cortex—the main region that was implicated in consciousness in the previous section.

STILL CONSCIOUS, BUT ONLY OF THINGS ON THE RIGHT

I began this book relating how my father temporarily had suffered hemispatial neglect following his stroke. This condition, associated either with damage to the prefrontal or parietal cortex in one hemisphere, has classically been the most explored attentional syndrome, though to my mind it is also the most central consciousness deficit there is—far more relevant to consciousness, for instance, than blindsight. Say the damage is on the right. Then, for that patient, for some of the time, the left half of space simply doesn't exist (the opposite side is affected because of the crossover wiring of the brain). It's not a deficit of vision, because if you do a sight test, these patients can see everything fine on both sides, in both eyes. Sometimes, also, if there is something particularly striking on the left, they will perceive it without any problem. But more often than not, these patients will just become distracted and fail to attend to anything in the left half of space. The classic test of this condition is simply to give patients a long horizontal line on a plain piece of paper and to ask them to put a vertical line in the middle of the horizontal one. Normal subjects have no trouble being very accurate at this admittedly very straightforward task. But neglect patients will almost always place a mark obviously closer to the right edge, in fact around three-quarters of the way to the right, just as if the left half of the line didn't exist.

Patients may also only shave the right half of their face, eat from only the right half of their plates, and, if asked to draw a clock face, would write all the numbers squashed up on the right half of the circle, or just leave the numbers 7 to 11 out of their drawing entirely. Similarly, if asked to draw a house, they'd merely draw the right half, with lines unfinished, half the windows missing, only half a door present, and a precarious right half of roof looking as if it could collapse at any moment. In every case, they are completely oblivious to the striking flaws in their behavior.

The condition extends to every aspect of the neglect patient's world. These patients will ignore sounds or touches on the left. And the disorder can even reach into their imaginations: In a simple yet striking experiment, Edoardo Bisiach and Claudio Luzzatti asked hemispatial-neglect patients to imagine they were in a square that was very familiar to them, the Piazza del Duomo, in Milan. They were first told to imagine that they were facing the cathedral from the opposite side and to describe all the buildings they could see. The patients reported all the buildings in their mind's eye, which were only those on the right of the square, from that perspective. They were then told to imagine they were on the opposite side of the square, facing outward from the cathedral, and describe all the buildings they could then see. They proceeded to fail to imagine or mention all the buildings they'd just listed a few minutes before, as these were now on their left in their mind's eye, but did describe all the buildings they'd previously failed to notice! So even in imagination, these patients were completely ignoring the left half of space.

The fact that you can either describe this condition as one of a profound attentional deficit or as an absence of awareness provides more ammunition for the view that attention is a critical component of awareness, functioning as its directional gatekeeper.

To reinforce the point that hemispatial neglect is not purely a visual impairment, Margarita Sarri and colleagues carried out an experiment on neglect patients in which she would touch either their right or left index finger while they were in the fMRI scanner. Half the time the patient wasn't aware of the touch on the left. The region of the patient's cortex responsible for finger sensations was activated regardless of whether the patient was aware of the touch or not, but if the patient *was* aware of the touch, then the intact portions of the prefrontal parietal network lit up.

Consciousness Shrinking to a Small Point

Habitually failing to be aware of half of your world is a profound enough impairment of consciousness, and yet there is an obvious question here—why aren't these patients more impaired, if these regions are so critical for consciousness? Why, for instance, can they occasionally still be largely aware of the left half of space? The answer lies in how much of the collective prefrontal parietal cortex is damaged in these patients.

In all the neurological cases I've described so far in this chapter, the damage has only included a small proportion of the total volume of the prefrontal and parietal cortices in both hemispheres. Most of the time the damage was only in one hemisphere, so at most would have only affected half of the volume of this network of regions. It's also likely that these brain areas, so flexible in function normally, would be more capable than any other region at adapting to the brain damage, with the intact parts of the network merely taking on more of the burden of processing. So what happens if the whole prefrontal and parietal network is damaged? Is such a person still conscious? Well, in humans I know of no case studies that have included such a profound loss of cortex—probably because such extensive damage would lead to death. However, there are rare reports of near complete damage to both prefrontal cortices or both posterior parietal cortices. Are these patients any more impaired than those I've described above? Unfortunately, they most certainly are.

The posterior parietal cortex, although in function highly similar to the prefrontal cortex, is marginally more linked to spatial processes. Patients who have extensive damage to both sides of the top portion of their posterior parietal cortex have a rare condition known as Bálint's syndrome, which completely knocks out their sense of space. Their experience of the spatial extent of the entire outside world seems to disappear, and patients report that "there is no there, there." One feature of this syndrome is that patients are unable to perceive more than one object at a time. Even with this single object, they cannot locate it, nor can they tell if it's moving toward or away from them. If they are looking at a crowded scene of multiple objects, once they recognize one object the others will become invisible. Sometimes, even perception of a single object is impaired. One patient, known as KE, could recognize colors and read words, but couldn't do both—so if he read a word, he was unable to identify its color. Although this condition is skewed toward the spatial domain, it nevertheless reflects the importance of the posterior parietal cortex for working memory. KE, and patients like him, have their spatial working memory effectively reduced to one—and in some cases not even a single complete object can be held in these patients' working memory—it will be stripped into its component parts, and only one of those parts will make its way into consciousness.

The prefrontal cortex is in some ways the most abstract general higher-order region that humans possess. Penfield's sister had most of her prefrontal cortex removed, but only in one hemisphere, with the left prefrontal cortex intact and able to take on functions that both prefrontal cortices had previously carried out. But if you lose both prefrontal cortices, what happens? Such cases are rare, possibly because the patient is far too ill to take part in any scientific research. But one neurologist, Bob Knight, has come across such a patient. The patient was awake, but otherwise looked disturbingly like a zombie. He had no motivation of his own, just constantly sat immobile in a chair as he stared into space, and tragically, it seemed as if any meaningful awareness was totally absent.

It seems clear from all these patients with varying degrees of damage to the prefrontal parietal network that such regions carry out a combination of attention and working memory processes, and that when such functions are severely impaired, consciousness consequently shrinks to a very marked degree.

Brain-Scanning the Prefrontal Parietal Network

But what is the consensus from brain-scanning studies exploring the role of the prefrontal parietal network? The main thrust of the research has indeed centered on working memory and attention, with the prefrontal parietal network closely associated with both.

For instance, the prefrontal parietal network will increase in activity if you increase the number of letters you place in working memory, the number of abstract relations between items of an IQ task you hold in working memory, and the number of spatial locations you have to remember. Likewise for attention. The prefrontal parietal network increases in activity, for instance, if you switch attention between tasks, or if you attend to visual changes on a screen.

In fact, as this is our most abstract, higher-order brain network, a common finding is that these areas are activated whenever we perform a complex or novel form of *any task whatsoever*, whether it involves short-term memory, long-term memory, mental arithmetic, or any other potentially challenging cognitive process. For this reason, these regions are also most closely linked with IQ.

This pattern of results implies two vital features of our thoughts and consciousness: first, the intermingling of cognitive processes, and second, how this amalgam of advanced mental functions is inextricably linked with awareness.

If you read a psychology textbook, even now, it will probably make neat distinctions between working memory, attention, long-term memory, mental arithmetic, reasoning, and so on, as though these were all independent processes. Increasingly, though, researchers are eroding the spaces between almost all types of thought and memory, and building broad links between them.

Aside from the fact that all such thoughts will activate the prefrontal parietal network, there are many other connections. For instance, it's clear that working memory is intimately linked with attention, with working memory effectively being the output of attentional filtering. Manipulations in working memory, especially of some current goal, can influence which attentional filters are set. And if you fill up working memory with items, your attentional ability suffers. To demonstrate this, Nikki Pratt and colleagues recently gave volunteers a classic attentional task that involved responding to the direction of a central arrow while ignoring a distracting group of surrounding arrows pointing the wrong way—a surprisingly difficult test requiring firm attentional control. On some of the trials, volunteers also had to keep a set of letters in working memory prior to stating which direction the central arrow was pointing. This extra working memory load caused the volunteers to guess the arrow's direction more slowly and less accurately—in other words, it reduced their attentional resources for the arrows. Pratt also found, using EEG, that brain-wave markers of attention for the arrows were weakened when working memory items were stored, further reinforcing the links between the two processes. Results such as these have led the most prominent current theories of attention to subsume working memory within an attentional framework.

Thus, this traditional picture that there are a group of largely independent brain areas and corresponding types of thought—say, one for attention, another for working memory, another for long-term memory encoding, and even another for consciousness—is one that increasingly clashes with the evidence. Instead, the most parsimonious way to describe the data is in terms of another distinction, that between static, automatic, unconscious processes, on the one hand, and highly dynamic, flexible, conscious ones, on the other. The automatic processes are those we've stored in our specialized memory and motor areas as overlearned habits and goals, and are usually the product of our roaming consciousness. And consciousness is an inextricably interlocked collection of processes carried out in the main by the prefrontal parietal network, with attention and working memory being the most prominent two functions. This consciousness machine is designed to kick in whenever the task cannot be achieved by our instincts or bank of unconscious automatic habits. It analyzes and manipulates the contents of our working memory in logical ways, drawing in further information from our specialist systems if necessary, and brings to bear this substantial portion of general-purpose cortex to solve complex or novel tasks, ideally by creating new automatic habits that the next time won't require consciousness.

THE PREFRONTAL PARIETAL NETWORK, CONSCIOUSNESS, AND CHUNKING

But what of chunking and spotting patterns? While it is indeed a near universal finding that the prefrontal parietal network will be increasingly activated as the demands of a task increase, there is one clear and important set of results with the exact opposite picture.

My colleagues and I at Cambridge carried out one of the first experiments explicitly to demonstrate this exception. Volunteers in the fMRI scanner would view an array of 16 red squares arranged in a 4 by 4 matrix. Four of these red squares would blink blue in a sequence, and a few seconds following this, the participants would try to correctly recall this sequence of locations. This is so far a standard spatial working memory test. The twist is in the fact that we actually gave volunteers two different types of sequences—ones that looked entirely random, just like the conventional spatial working memory tests, and another that took advantage of the 4 by 4 structured array to make the sequence form squares, triangles, or other symmetrical and regular paths (see Figure 8). This made the sequences easily chunkable. Subjects consciously detected these chunks and talked after the experiment about how these sequences were easier because of the patterns.

If difficulty in every single situation drives the prefrontal parietal network, then the unstructured, more difficult sequences should generate more activation in these regions. But if instead there is something special about chunking processes, then perhaps these easier structured sequences will drive the prefrontal parietal network more than the harder, unstructured alternatives.

In fact, very robustly, the easier patterned path trials did light up the prefrontal parietal network more brightly compared with the more difficult, unstructured sequences. So in some cases, at least, task demands don't drive this network—at least when chunking is involved.

Because this was a rather unexpected result, we repeated the same experiment, except this time with digits: Subjects heard 8 single digits in the fMRI scanner, and then after a few seconds had to say back the sequence in the order that they'd heard it. Some sequences were distinctly structured, such as 8 6 4 2 9 7 5 3 (descending even numbers, then descending odd numbers), while other sequences were deliberately made to be as random as possible. Just as with the spatial structured and unstructured trials, these structured digit sequences were easier because subjects could chunk them—and they, too, activated the prefrontal parietal network more.

One outstanding question we had with this work was whether it was simply memory for preexisting chunks, such as squares or single-digit, even-number sequences, to which we are all exposed from early childhood—that was driving the participants' responses, or if participants were instead noticing the chunks on the fly, spotting the pattern as a powerful new rule to apply on each trial. So we devised a new experiment, sticking with verbal working memory and digits, but this time moving on to sequences of double digits. This way we could give volunteers sequences they were very unlikely to have seen before, to make sure we were dealing with mathematical and novel pattern spotting. For instance, subjects might have to remember the sequence 57 68 79 90 over a few seconds (each number going up by 11 each time). We also had a standard, totally unstructured sequence that subjects couldn't apply any chunking strategies to (say, 31 24 89 65). And there was also a memory-based chunking condition, to see if that would also activate the prefrontal parietal network, and if so, how strongly. To enable a memory-based chunking condition, I trained my volunteers for at least 4 hours each to memorize 20 different unstructured 4-digit chunks before I scanned them. I told them they had to imagine they were playing a game where they were a new receptionist at a medium-sized company, and as part of their induction they had to memorize the faces, names, and phone extension numbers of 20 key members of staff. So, after a series of graded training exercises, they each had a relatively naturalistic set of 4-digit numbers cemented in their memories. When it came to the scanning part of the study, they might have seen the novel sequence 21 05 81 63 to recall in working memory, where 2105 would be one phone extension number they'd comprehensively learned during the previous week, and 8163 would be another. So we had three different fMRI digit-based working memory conditions—one involving mathematical structure, a second involving mnemonic structure (made up of the phone extension numbers they had learned so thoroughly), and a third with no structure (so subjects just had to rely on their working memory and not chunking). With these three, we could distinguish the brain regions activated for mathematical chunking against those for memory-based chunking. In fact, we also had two controls—one to match the memory content and a second to match the mental arithmetic content, but in each case without the additional digit sequences task, and thus the opportunity to benefit from chunking.

As usual, the sequences that could be chunked, because of strategies involving either memory or mathematics, were easier for the subjects to recall compared with the unstructured sequences. And both kinds of chunking sequences, as before, lit up the prefrontal parietal network considerably more brightly than did the unstructured sequences, where chunking wasn't an option. Additionally, the condition involving mathematical sequences still robustly activated the prefrontal parietal network compared to its matched control condition (where subjects carried out equivalent mental arithmetic tasks, but without any chunking component). A similar pattern of prefrontal parietal activity was seen when the condition involving the memory-based chunking sequences was compared to its own matched control (with the same level of memory recall, but no chunking aspect to the task). This demonstrated that these regions were not just being driven by mental arithmetic or memory recall when subjects carried out the chunkable sequences, but something more—quite specific to the act of chunking. I was most struck, though, that the condition involving mathematical chunks *still* robustly activated the prefrontal parietal network when compared to the memory-based chunking condition. This placed the mathematical chunking task as the firm winner in the game of driving activity in the prefrontal parietal network. Ask a scientist in the field to generate a list of those processes that will most reliably activate the prefrontal parietal network, and she will most likely include working memory, long-term memory, and mental arithmetic. But in this experiment, the mathematical chunking task activated the prefrontal parietal network more strongly than all of these, as well as compared to the memory-based chunking condition. In other words, this experiment demonstrated that the prefrontal parietal network will activate for many complex tasks, but it will be most excited when subjects are actively searching for and finding entirely new patterns.

Other research groups have also linked chunking with the prefrontal parietal network. In the long-term memory domain, Cary Savage and colleagues showed that chunking will both light up the prefrontal cortex and boost performance if a group category strategy is applied to word-list memorization (for instance, remembering all plants together in one group and all metals in another). Similar results have been found in the working memory field. For instance, Vivek Prabhakaran and his coworkers presented letters to participants with the instructions that these stimuli had to be recalled over a short delay. If volunteers chunked the letters using a strategy that involved binding a spatial location to each letter, then prefrontal activity increased. In another working memory task, Christopher Moore, Michael Cohen, and Charan Ranganath demonstrated that extensive training involving the categorization of abstract shapes enabled memory-based chunking for such stimuli, which improved subjects' performance and increased activity in the prefrontal parietal network.

One recent intriguing, related study by Stanislas Dehaene and his team actually showed the transition from consciously spotting the pattern to then using it in a routine, automatic manner. Subjects had to discover a novel sequence from the letters ABCD. For instance, a participant might first try "A" and be told this was wrong; then she might try "B" and be informed that this was correct—so now she would know that the first letter in the sequence was "B." She might try "C" for the second letter and be given feedback that this was wrong. At this point, she would have to start the sequence all over again, but at least she would now know to start with "B" and explore the second letter. Eventually, by trial and error, she would work out the sequence, which she would have to repeat between three and six times—now a very easy task for her. Then there would be a new sequence for her to work out, and the cycle would continue. During the initial search phase for each trial, there is massive prefrontal parietal network activity, but this quickly dies down as soon as the task becomes routine. The prefrontal parietal network then becomes virtually silent as the task is automated in the volunteers' minds, and they need very little consciousness to complete the well-learned sequence. In other words, this is a beautiful illustration of the distinction between the conscious search for

patterns, carried out in the prefrontal parietal network, and our largely unconscious automatic habits, which require specialist brain areas alone.

In all the examples above, the chunks tended to be so obvious that volunteers
couldn't help but be aware of them. What would happen, though, if the participant couldn't consciously spot the chunks for some reason? Would the prefrontal parietal network fail to activate for these undetected structured
sequences? Serendipitously, I was able to answer this question via a fascinating
person who was effectively blind to all such number chunks. Daniel Tammet
is a prodigy with a few similarities to the incredible Russian mnemonist Sherashevski, discussed in Chapter 2. For one thing, Tammet has a rather extreme
form of synesthesia, just as Sherashevski did. Those with synesthesia commonly
associate colors with specific single digits they read. Tammet, however, has a
different experience for not just the first ten numbers, but the first ten thousand.
Not only this, but he also adds not just color, but texture, shape, height, and
even touch to his perception of numbers. This creates an incredibly vivid, structured experience for him whenever he reads a stream of digits. He is also diag-
nosed with a high-functioning form of autism known as Asperger's syndrome.

When I tested him, Tammet was able to cram far more numbers into his
short-term memory than any other volunteer I'd ever tested (though this was
somewhat reduced when I deliberately confused him by coloring the numbers
in ways that clashed with his synesthetic inner eye). He had just set the European record for memorizing the most decimal places of pi: 22,514—which
he claimed was quite easy to do, the most difficult part simply being the 5-
hour stint needed to recite all the numbers from memory. He also has prodigious mental arithmetic skills—for instance, being able to divide one double
digit by another and give the answer to 100 decimal places. And he claims a
deep facility for languages, with the ability to learn a new language in a single
week. Although his Asperger's syndrome probably allows him to concentrate
more deeply than most, his exceptional abilities mainly arise from the intensely multisensory inner numerical world he experiences, with every number seeming so very vivid and distinct to him. Memorizing or mentally
manipulating numbers comes easily and naturally, and to recall them, he simply has to convert the inner psychedelic mountain terrain back into digits.

We decided to investigate his brain activity when he carried out one of
our chunking tests—where 8 single digits are presented to be retained in

memory over a few seconds, with half the sequences structured, like 8 6 4 2 9 7 5 3, and half random and unstructured. When I asked him after the experiment whether any trials were easier than any others, he unsurprisingly said that they all were just as easy—because for him, unlike normal people, remembering only 8 digits really is very straightforward. But when I probed further, it turned out that, surprisingly, he'd completely failed to realize that some of the trials involved highly structured sequences of digits. He was in fact completely blind to the external structure. His brain activity reflected this: Totally unlike normal volunteers, he showed absolutely no increase in activity for the structured sequences compared with the unstructured sequences. He was neither aware of the structures nor in any way exploiting these patterns to chunk the sequences and reduce his working memory load. So Tammet showed, by failing to notice these chunks, and failing additionally to activate his prefrontal parietal network for these obvious forms of external structure, that you really do need to be aware of and use the chunk in order for the prefrontal parietal network to kick into action.

But for Tammet, it certainly wasn't as if absolutely no structure was imposed on these sequences. He might not have noticed any additional structure for the mathematically chunkable sequences, but for him, because of his rich form of synesthesia, in a sense every trial was highly structured—it's just that the structure resided in his head, in the multisensory mountain range of his mind's eye that appears whenever he thinks of a stream of numbers. Reflecting this, for all trials, on average, whether they were structured or not, he had markedly increased activity in the prefrontal part of his prefrontal parietal network compared to the normal volunteers. This is because, for him, every trial, not just the structured trials, were being chunked at least to some extent. So, in two unexpected ways, this experiment reconfirmed the links between chunking, consciousness, and the prefrontal parietal network.

This collective evidence, then, shows that consciousness is most closely connected with the prefrontal parietal network, which supports not only attention and working memory processes but also any kind of novel or complex task. But if you want to activate this network the most powerfully, and by extension engage your consciousness to its most intense heights, you need to detect some useful pattern. These chunking processes are an embedded part of our advanced conscious cognitive machinery. They are also perhaps the essence of what it means to be conscious. They are the mechanisms by which we convert awkward obstacles into innovative solutions and initial, error-prone fumblings into adept automatic habits.

HARMONIOUS EXPERIENCES

Admittedly, though, these are broad strokes in painting the details of how the brain creates consciousness. One way of looking at the problem in a deeper way is to ask how neurons communicate with each other in order to generate our experiences. Although this question was partially answered in the previous chapter, by my description of how attention equates to coalitions of neurons competing for dominance, another feature of neuronal communication is the waves of activity used to connect brain cells together. The main tool for examining this neural chatter is EEG, which lacks the fine-grained spatial resolution of fMRI but can collect data every millisecond, compared to the second or two it takes for an fMRI scanner to grab a picture of the brain's activity.

Francis Crick was the main early champion in this area. One of the most famous scientists of the twentieth century, Crick codiscovered the structure of DNA in the early 1950s, and proceeded to contribute revolutionary findings in genetics for a generation before turning in the last two decades of his life to the science of consciousness, which he believed was the greatest unsolved mystery in biology.

Crick popularized the idea that when neurons act in harmony in a certain way, then consciousness ensues. The particular frequency bandied about for this love-in between neurons was originally the gamma band, roughly averaging to 40 cycles per second, one of the fastest frequencies that neuronal communication is capable of (and a frequency previously linked with attention).

Although there is solid support for the link between this gamma band and awareness, this thesis requires a couple of tweaks. For instance, in rats at least, these fast waves are observed, along with slow delta waves, when the animals are under general anesthesia. Rats can also generate these swift types of neuronal synchrony when in a deep sleep.

The resultant updates to this suggestion that the gamma band reflects consciousness suggest that local communication between neighboring neurons using these frequencies is not sufficient for consciousness to arise.

What's required is a gamma rhythm binding together the information between neurons that may be on separate neural continents—for instance, one region in the prefrontal cortex toward the front, and another in the posterior parietal cortex toward the back.

And perhaps gamma isn't quite fast enough for consciousness—instead, what's called "high gamma," with frequencies from 50 cycles a second, possibly even up to 250 cycles a second, is currently a hot topic in consciousness research. This frequency band is too swift to study using conventional EEG because of the interference of the signal by the scalp. But some epileptic patients soon to undergo a brain operation to remove the locus of their epilepsy have EEG arrays implanted directly onto their cortex, under their scalp, in order to investigate where they are having their seizures. When the electrodes are immediately over neurons, then you can pick up such high frequencies with ease. Two independent labs, Stanislas Dehaene's near Paris and Bob Knight's at the University of California at Berkeley, have both shown, using this technique, that when the patients are free from seizures and normally awake, these ultrahigh waves of neuronal activity are exquisitely connected with consciousness, and might, in concert with and locked to lower frequencies, be the main neuronal signature of awareness. Why such high frequencies?

The answer follows from the purpose of awareness. In order for consciousness to carry out such complex tasks and generate insights, it needs to make connections in two ways: First, it needs attention, not just to select the most pertinent mental objects to place in working memory, but for all aspects of that object to be knitted together into a single, coherent whole. So when I see Angelina Jolie in that red dress on the silver screen, I don't see her eyes as one object, her nose as a separate, disconnected object, her hands and name and voice as others. Instead, she is just a single, unified, though complex object, with many different components all connected together. Very importantly, I really can't help consciously recognizing her *as Angelina Jolie,* as opposed to all these distinct parts.

The color of her dress may nevertheless be represented in my visual cortex at the back of my brain, her face in my fusiform face area at the bottom of my cortex, her name and various other facts about her in my semantic store, at the front of my temporal lobes, and so on. When I recognize Angelina Jolie, many specialist cortical areas, all over my brain, are involved in that recognition, as well as my prefrontal parietal network, which acts as a conscious, temporary manager of the information in working memory. Now, if all of these spatially disparate regions need to be joined together to represent the conscious mental object, Angelina Jolie, a slow neuronal rhythm, say of a few cycles a second, simply isn't up to the task, since there is too much data to hold together in the time. This slow kind of rhythm, the delta band, is instead commonly found when someone is under general anesthesia. The conscious frequency, in contrast, is as fast as our brains can allow. High gamma waves of activity are initiated by the thalamus, the Grand Central Station of brain regions in the center, and then these ultrafast waves perpetuate through the relevant parts of the cortex and bind together all components of an object in consciousness.

The second type of connection required for consciousness is between those items in working memory in order to spot patterns or chunks, or simply to maintain a sequence of items. A fast rhythm between neurons throughout the cortex can maintain the links between distinct objects and allows us to analyze and manipulate them in working memory.

EEG, with its millisecond accuracy, is also excellent at providing a time frame for consciousness. These widespread high gamma rhythms don't immediately follow the presentation of a stimulus. Instead, they are relatively protracted, as you might expect when much of the brain has to coordinate its activity, and they take at least 300 milliseconds to form. This is a very similar value to the time it takes attention to filter sensory input according to our current goals.

Consciousness, in Theory

I can now paint an overall empirical picture of consciousness. If a vivid red rose comes into view, my experience of it is built up over a third of a second, as an initially brutal neuronal competition leads to the shaping of brain activity around my attention toward the rose. An ultrafast, harmonious neuronal rhythm spreads outward from the thalamus and merges my collective neural information of the rose, which is stored in specialist areas throughout my cortex. This high-frequency, long-range, unified mental chunk will also broadcast itself into the prefrontal parietal network, where the experience will come to life.

But if I were faced with a more novel or complex task, my consciousness would show its true potential. My prefrontal parietal network activity would reflect an engaged working memory, a focused attention, and a ravenous search for patterns in order to conquer whatever mental obstacle was in my way. Meanwhile, my specialist regions of cortex—for example, areas that store knowledge about objects at the front of the temporal lobes—would take turns to support my consciousness by providing the specific contents to my experiences.

The next challenge for consciousness researchers, in the decades to come, is to discover exactly how neurons collectively represent the information they do and what forms of neuronal interaction generate consciousness. For instance, just how is information transmitted between the fusiform face area and the prefrontal parietal network, via these high gamma waves, to generate my experience of my daughter's face? And what is the precise code that neurons use to represent information? Such questions may be answered by simultaneously recording the activity of each of thousands of neurons in multiple regions. At present, the state of the art is limited to dozens of simultaneous electrodes (in the macaque monkey, the closest species to us where these studies are routinely carried out), so the technology falls considerably short of the kind of data collection required. But there's every reason to assume that in the next decade or two the methods will be sufficiently advanced for us to discover and extract the precise neural signature of consciousness.

In the meantime, many scientists have created detailed theories based on the existing empirical picture. Admittedly, there was a rather wild crop of early theories in the final decade or two of the twentieth century—for instance, one seemed to rely on the impeccably argued syllogism that because consciousness is mysterious and quantum mechanics is mysterious, then quantum mechanics must explain consciousness. But now the story is very different, with theories linking closely with the latest empirical findings. What is striking about the most prominent current crop of theories is how they are all broadly converging on the same overall position.

The three most popular serious theories of the day all, at their heart, see consciousness as a particular flavor of dense information transmission across a large cortical network. But each theory differs subtly in its particular perspective on this general view.

Victor Lamme's *recurrent processing model* starts with the stark assumption that we may think we know when we are conscious of something, but we couldn't be more wrong. According to Lamme, there are many times when we are actually conscious but we don't even realize it. So he abandons talking about psychology, and what experiences we can or cannot report, and so on, and instead centers entirely on what is happening in the brain. Sometimes one brain region will feed information to another, but the second brain region won't talk back to the first. Other times, there will be "recurrent processing," where both brain regions are entering into a proper back-and-forth, two-way dialogue as they exchange information. Lamme believes that it is only when this second kind of neuronal chatter is taking place, with information bouncing between regions, that consciousness occurs. If this back-and-forth talk happens only between specialist areas, such as different visual cortical regions, then there will be some level of consciousness, but it will not be strong enough that we could say, for instance, "Ah, there's a red rose in front of me." But if this two-way communication stretches into the prefrontal parietal network, then we have a full, deep consciousness and can report on what we're seeing.

Lamme's notion that recurrent processing is required for conscious levels of information transmission to take place is a very plausible suggestion. But I find his insistence that we are still conscious even when we are quite convinced that we are not to be a deeply unpalatable stance. In order to build a coherent theory of consciousness, it's fine to be suspicious of the edges of what we report about our experiences, but it is not sensible completely to ignore the very event you are trying to explain. Partly because of Lamme's rejection of the experiential intricacies that make up how we are aware of the world, his model fails to capture much detail about the nature or purpose of consciousness.

The model most closely aligned with the existing data, and the view of consciousness I've been describing throughout this book, is the *global neuronal workspace model* proposed by Stanislas Dehaene and Jean-Pierre Changeux. This model is largely the neuronal extension of the global workspace theory put forward by Bernard Baars. In Baars' theory, consciousness is roughly equated with working memory. It's a spotlight on a stage, or scribbles on a general-purpose cognitive white board, which lasts only a second or two, but which can contain and manipulate working memory items by drawing them from our vast unconscious reservoirs of knowledge in specialist nonconscious systems.

In the global neuronal workspace model, the brain also divides along specialist and generalist lines. First there are the specialist, content-dedicated areas at the edge of the collective brain network, which store our memories, crunch data from our senses, and so on. The neurons here have short to medium connections with each other and are all that's needed when we perform an effortless, automatic, largely unconscious task. The inferotemporal cortex, processing visual objects, is one example of such a region. Then there are the general-purpose regions at the densest center of the network, comprising the prefrontal parietal network and thalamus, which "ignite" in dramatic activity whenever an effortful task is required, so that this entire core can become fully activated simultaneously. This central core includes lots of long-range connections between neurons, enabling this neuronal workspace to draw in specialist knowledge from the content-dedicated regions at the thinner outer edges of the network. If necessary, the prefrontal parietal network and thalamus can also control and modify the activity of these subordinate distant areas, so that complex information processing can occur and difficult tasks can be achieved. Activity in this core set of regions, particularly involving the prefrontal parietal network, is the locus of consciousness.

Anatomical studies of how the brain is wired put the lateral prefrontal cortex, one of the main regions of the prefrontal parietal network, at the top of the league in terms of how many other regions it is connected to, although the posterior parietal cortex and thalamus are not far behind. So in terms of brain wiring, the prefrontal parietal network, in concert with the thalamus, constitutes an "inner core" of regions that are ideally suited to collect information from the rest of the brain, carry out the most complex tasks we are capable of performing, and generate our sense of experience from this central hub of information processing.

But because Dehaene's model is so closely wedded to the empirical neural details that are associated with awareness, he has opened himself up to the charge of not more ambitiously capturing the mechanistic essence of consciousness.

The third and final theory, Giulio Tononi's *information integration theory*, travels in the opposite direction, only discussing mechanisms while refusing to get its hands dirty with too many tawdry details about the brain. Information integration theory is also by far the most abstract and ambitious of the current crop of consciousness models.* An entirely mathematical theory,

---

*This theory is similar to two other modern theories of the brain, proposed by Gerald Edelman and Anil Seth, which also try to mathematically equate consciousness with the complexity of information the brain processes. I chose to highlight information integration theory because it is the most prominent and detailed of the three.

it tries to distill consciousness into its informational essence. Whereas the previous two models were rooted in the real cortical networks of the human brain, Tononi's theory applies to any network of nodes whatsoever, be they a connected series of neurons, computer transistors, or any other information-carrying object one would care to imagine. For Tononi, a network's capacity for consciousness is directly related to how many different kinds of information it can represent and how well those pieces of information can be combined. The more nodes there are in a network—as long as they are sufficiently connected to each other—the more varied the possibilities for combined forms of information and the greater that network's capacity for consciousness.*

This simple yet powerful recipe for awareness can map quite neatly onto concepts such as attention, with its drive to combine information about an object into a unified whole, or even the previously mentioned global neuronal workspace model, which also cares about a dense central network to carry combined forms of information.

Under the information integration theory, regions like the cerebellum can never support much consciousness, as they have few connecting internal wires. Specialist brain regions, such as the primary visual cortex, can play only a minimal role in consciousness, again being at the edges of the main network. In contrast, the prefrontal parietal network, being so densely interconnected and also linked to many specialist regions, is just the kind of network shape that can support high levels of consciousness.

Without the prefrontal parietal network, we are really just processing an independent collection of facts in parallel—for instance, the color of Angelina Jolie's dress as one datum, the sound of her voice as a completely separate feature, and so on. But when there is some highly interconnected

---

*Actually, in this model there definitely is such a thing as too many connections, and some ugly compromise of connectivity is the ideal. The critical factor here is just how many different states a network can be in. This can be physically imagined by appeal to symmetries. Say there is a 3 by 3 by 3 cube of nodes, thus 27 nodes total. If all the nodes are connected to all the others (or similarly, if none of the nodes are connected), one corner node lighting up is the same as all the others. But say there is a middle ground, with each node connected to somewhere between 5 and 20 of the other 26 nodes. This time when a corner node lights up, its connections mean it is unique; there are no other corner nodes with the same configuration of connections as this one, and so its information state is unique. Because there are no symmetries of wiring, this particular cube could represent the most information out of the three options.

network involved, such as the prefrontal parietal network, with attention combining those facts within it, the richness of that information far exceeds the sum of each individual piece of data, and consciousness ensues.

And just how much information of this merged form a network can contain is the same as how many different states of activity it can be in. In practical terms, this means how many different kinds of experiences we are ever capable of. I might have inferior senses to a dog, and therefore at best a matched level of information input, but because I can combine the data from my senses in so many different ways, due to the powerful analytical machine of my prefrontal parietal network, the range of experiences I can have far exceeds that of a dog.

Two questions concerning our inner mental life have, for centuries, obsessed philosophers: (1) Why is consciousness inherently subjective? and (2) How can a physical lump of matter give rise to the glorious variety of sensations we can experience, from seeing a red rose to hearing a Beethoven symphony? Tononi ambitiously claims to be able to address both of these issues.

For Tononi, the inherent subjectivity of awareness, where my experiences are private to me alone, impenetrable to anyone else, is an essential component of consciousness, because that is precisely what the mathematics of his theory predict. If consciousness is just the activity in the densest part of the network of my brain, how could my consciousness extend outside my brain—for instance, to another person? There are no network connections to make this possible, for a start. But even if these connections somehow were present—for instance, I had some sort of neural graft connecting my brain with another's—if this connection was too weak, then there would be two dense networks and two consciousnesses, with a very weak experiential connection between the two. This situation would be broadly similar to that of the conjoined twins Tatiana and Krista, who are very much two different people, but who just happen to share the occasional feeling. So subjectivity, far from a philosophical conundrum, might simply be a product of the way that closed networks generate their compound, unified items of information—a mathematical, computational process that we happen to call consciousness.

And what imbues my awareness with all the different sensations that I can experience? Why do my baby daughter's dark brown eyes appear that particular color to me? Why does her sweet, high babbling voice have that precise sound? And why does my stroking her soft cheek as I send her to a night of sleep have that particular feel? According to Tononi, whatever we experience is permeated by its specific sensation because of the precise point it takes up in this huge space of possible forms of combined information we can represent. The dark brown shade of my daughter's eyes feels like it does because of the informational contrast it makes with all the other colors I could possibly see, which evoke similar experiences for their similar informational content. But that particular experience of color is also in contrast to all the other experiences I could have in my other senses and beyond, which feel less similar because of their greater informational distance. So every experience we could possibly have, as a unique set of pooled information, gets its distinct perceptual characteristics from its relationship to all our other possible experiences.

Some might argue that these bold claims for consciousness are similar in validity—or lack of it—to any opposing philosophical claim. I think Tononi himself would be the first to admit that his theory, mathematically dense though it is, currently exists only as a skeleton of a framework for consciousness. And there are rather strange corollaries of the information integration theory in its present form that sit rather uneasily with much of neuroscientific data. For instance, the theory doesn't in any way equate a heightened neural activity to increased consciousness, so from the perspective of the theory, a quiescent brain is just as likely to produce a highly conscious state. This clashes with virtually all the studies I've discussed throughout this chapter. Moreover, there is no distinction in information integration theory between useless, unpatterned forms of information and highly structured, profound chunks of insight, which I've repeatedly argued are really the hallmarks of consciousness.

But what Tononi has done, quite differently from the philosophers, is to provide a very brave, thought-provoking first sketch by which to explore deep questions about awareness in a scientific way. Because information integration theory allows you to compute a single number for just how much consciousness at a given time resides in an animal (or robot, for that matter), based on its network architecture and changing states, many opportunities for exciting experiments immediately arise. In practice, though, in the present form of the theory, this "consciousness number" carries too great a calculation burden for today's most advanced computers, even for simple sim-ulations of animals with only a handful of neurons. So there are, at present, severe practical limitations to information integration theory (although I will be discussing in future chapters the methods scientists are using to make useful approximations to this magic number).

In fact, a common drawback of all the neural theories of consciousness, one way or another, is that they tend to focus on mathematical or neuro-physiological details while somewhat neglecting the psychological compo-nents that make up consciousness. In my view, the next iteration of these models will need to place attention as a key ingredient in the recipe for con-sciousness, rather than assume it is a separate process, as many scientists currently seem to do. And if they include working memory at all (at present only the global neuronal workspace theory does), then they should flesh out the details of this—such as including chunking processes as a fundamental component and explaining just how our consciousness is so ripe for inno-vative thoughts.

Ultimately, the test of these models will be whether they can predict a surprising psychological feature of consciousness, as well as its correspon-ding neural details, that we can then confirm in a future experiment. This current set of theories, despite their great promise and encouraging conver-gence, are a few stages removed from this predictive level as yet.

Explaining Experiences

Even if there are still small but significant gaps in consciousness research and theory, nevertheless an exciting picture is emerging. This perspective, for the first time, provides a coherent explanation for the evolutionary ori-gins and purpose of consciousness, as well as the mental architecture and neural underpinnings of our experiences.

We all have learned at school that for life to exist you need a suite of cer-tain chemicals, measured out in the right proportions—water, carbon, and so on. And such ingredients are indeed essential. But what is far more fun-damental—what matters, above all—is that in this blind struggle for survival and reproduction, in this delicate dance between genes and the environ-ment, there is a pressure to gather information. All organisms are biological computers, forging a successful niche in the world by iteratively gathering a specific bank of useful, implicit knowledge. All successful life, via the ideas stored in its DNA, shapes itself to the relevant features of the world.

But animals, additionally, can learn more dynamically. They can gather vital new facts to help them survive and adapt, not just between generations, but within their own lifetimes. At birth, we humans are terribly vulnerable, with few instincts and poor senses, but with a mind wide open. In our lifetimes, we absorb a staggering quantity and variety of information. Although simple features of the world can be detected below our conscious radar, we funnel the most useful, novel, and complex information, via attention, into the tightly limited, but also unimaginably flexible, workspace of our conscious minds. In this playground of possible experiences and ideas, we spot insights, we see the hidden regularities in the world, and, in so doing, we understand and comprehensively conquer the planet. It is our consciousness that is responsible for this—a consciousness that builds pyramids of information that we readily retrieve and manipulate. Our brains are exquisitely, densely wired so that information can easily flow from specialist regions, which store our refined gems of knowledge, into other areas that have no specific role themselves, but are so thickly connected with the rest of the brain that information, probably via high-frequency waves of neural activity, can easily flow and combine. Here, in the intensive neuronal murmuring of the thalamus and prefrontal parietal network, we do so much more than link raw redness with a simple shape; we experience the sight of a vivid crimson rose, we smell its soft aroma, and connect it with the exciting new romantic flame we just handed it to, as we anticipate many future experiences to share with her.

Now, with the psychological and neural features of consciousness fleshed out, I'll turn to how this picture of our experiences can be applied, first to assessing awareness in those beings that lack the language to tell us. I will then turn to the clinical implications of this new scientific view of our experiences.

6

Being Bird-Brained Is Not an Insult

Uncovering Alien Consciousness

Tender Chimps, Capricious Bonobos

Monkey World, in Dorset, England, is the largest primate rescue center in the world. Although it looks after orangutans, gibbons, monkeys, and lemurs, primarily the center houses chimpanzees, with the highest number in any one place outside of Africa, currently about sixty strong. This group of chimps, largely saved from neglect or abuse, is cared for by thoughtful, devoted staff, and the animals seem to lead a relatively happy, natural life. Monkey World is such a popular location that multiple television documentary series have been made describing the fascinating activities at the center. The following story is taken from one episode.

Some years back, Olympia, like all female chimps in the center, was given a birth control implant, but managed somehow to remove it, and promptly became pregnant. She was a devoted mother for a year to her daughter, Hebe, but then her milk dried up. The Monkey World staff made the difficult decision to remove Hebe from her mother for a while so that a human could continue to feed her. Six weeks later, they started reintroducing Hebe to her mother in short visits, with a view eventually to returning Hebe to Olympia full time, when food was no longer an issue. Jim and Alison Cronin, the husband-and-wife team in charge of Monkey World at the time (sadly, Jim Cronin has since died), were concerned about how the mother and daughter would react, so decided to take very tentative steps toward a reunion. Their most experienced primate manager, Jeremy Keeling, moved into a special nursery room with Hebe, who was clutching tightly to Jeremy with all four limbs when her mother, Olympia, was brought in.

Olympia clung impatiently to the bars of the door, knowing her daughter was on the other side. As soon as the door swung open, Olympia walked in and maneuvered herself to be ready to take Hebe off her human caregiver, Jeremy, and excitedly give her a hug. But Hebe wasn't so sure who this adult chimp was and became distressed, clinging ever more tightly to Jeremy. Olympia didn't get frustrated, though. Instead, she tried to innovate. She played it cool. She walked away and started to make a nice inviting, cozy nest out of the straw in the far corner of the room. Hebe slowly realized that this big chimp wasn't a threat and maybe, possibly, was someone special in her life. She left Jeremy's safe human hands and walked a few steps toward the nest and her mother's arms. But as Olympia gently approached, Hebe had second thoughts, backing up again toward Jeremy.

So Olympia tried another tack: Hebe and Jeremy were sitting on a table, so Olympia waited under it and held out her arms invitingly, ready to catch Hebe if she fancied a fun jump toward her. Hebe still wasn't convinced. The next strategy was for Olympia to pretend to ignore Hebe and instead groom Jeremy, picking intently at his thick moustache. Finally, Hebe was relaxed enough to allow her mother gently to touch her arm. Carefully Olympia looked all over Hebe's body, as if to check that she was all right. Progress— but still not the hug between mother and baby that Olympia sorely wanted. So Olympia tried one last thing: She pretended to leave. She walked out of the room, opened a metal cage door, and shut it again from the other side, all the time watching her baby. This—almost—had the desired effect. Hebe left Jeremy and rapidly approached the door and her mother, as if she didn't want her to go. Olympia quickly opened the door and held out her arms for a hug, but Hebe's nerves returned, and she rushed back to Jeremy. The staff called it a day for this first attempt, and they left a distressed mother to ponder things over with her main group, which soon came to join her. The next attempt, a day or two later, was far more successful.

It's possible that everything that happened in the above scene was behavior with no conscious life at all. It's also possible that every thought and feeling we project onto these chimps is real—that a mother, who deeply missed and loved her daughter, calmly understood that this daughter no longer recognized her. So she put herself firmly inside the mind of her young, frightened child and tried anything she could think of tenderly to calm the baby and once again make her embracing arms an inviting presence.

Bonobos, a very close relative of chimpanzees, and at least as intelligent, also exhibit behaviors that appear to reflect a burgeoning consciousness highly comparable to our own. Sue Savage-Rumbaugh, who investigates the extent to which bonobos can learn language, in her book *Kanzi*, coauthored with Roger Lewin, recalled the following story about a mature female bonobo named Matata:

> Once, when I was introducing a new person to Matata, she became jealous of the new person and refused to let her touch any item that she was fond of, including her blankets, her bowls, her food, and her mirror. One day we were sitting together on the floor when Matata decided to ask me to go get some food by holding her empty bowl out to me and making a food sound. I told Matata I would get her some food and left the room, leaving the bowl with her. I had been gone less than five minutes when suddenly Matata began screaming loudly. When I rushed back into the room the new person was holding Matata's bowl and Matata was screaming at her and threatening to bite. Matata looked back and forth from me to this person and then to the bowl, screaming—intimating that the bowl had been grabbed from her in my absence and that I should support her in attacking this mean individual who had taken her bowl. Of course, given the gift of language, this person was able to explain that she had indeed done nothing. Matata had placed the bowl in her hands and then started screaming for me as if she had been wronged.
>
> When Matata saw us talking about what had happened, she began to look very crestfallen, concluding her ruse had not worked. She stopped screaming and moved to the corner where she suddenly became very preoccupied with grooming herself.

The above two examples, at opposite poles of complex behavior—one tender, the other capricious and possessive—provide tantalizing anecdotal evidence that there are nonhuman species that lead a complex conscious life not that dissimilar from our own. But how can we establish that this is true in our closest evolutionary cousins, let alone in far simpler or more alien creatures?

In this chapter I'll be describing how we can make headway in assessing consciousness in all these nonlinguistic beings, with animals as the main model. I will also discuss the common intuition that even if other animals are conscious, there is something quite special and unique about the kind of awareness that humans possess.

I will be outlining two main angles of attack here: First, an analysis of those forms of animal behavior that seem most redolent of consciousness, if the same behavior were seen in humans; and second, what types of brain structures and functions are the most suggestive of consciousness in any life-form.

CRAFTY CROWS

But how does one go about assessing whether pigs, for instance, are con-scious. Outward signs that they are in distress—squealing when they are in pain—might be a hint that they are aware. But a skeptic will claim that this could purely be a behavioral response, that the animal was programmed to make that noise when it had a certain pressure applied to its skin, but that there is not necessarily any conscious life in between the pressure and the squealing. The obvious yet important fact that animals can't use language to tell us whether they are aware means that, technically, the skeptic's position is an entirely valid one.

A slightly more suggestive feature of animal behavior than suffering, often overlooked in discussions of the extent of their awareness, is that many animals appear to get bored. Humans may well be ravenous for information, but many other species, too, seem to have a hunger for knowledge that they critically need to meet. If you place animals in a drab lab cage, a dark, simple farm pen, or a zoo enclosure with no objects with which to interact, many of them will soon develop stress behaviors. They will pace up and down, rock back and forth, or simply start to appear apathetic, spending much of the day sleeping and resting. If left for long enough, many animals show signs reminiscent of human self-harm or outright mental illness: A parrot may peck off all its feathers, a rat may nibble off the ears of its offspring, monkeys may try to gnaw off their own limbs. These signs of abuse can be at least as disturbing as if the animal were regularly beaten.

The extent of suffering in these cases is particularly tendentious, since it arises out of a poverty of intellectual stimulation. But perhaps more direct evidence of animal consciousness can come from the extent of animal intelligence. Here, we needn't necessarily look to obvious examples like our great ape cousins to see signs of impressive mental capabilities. Nicky Clayton has devoted her career to studying corvids, a bird family including crows, ravens, jays, and magpies. Along with parrots, they are effectively the geniuses of the bird world, with large brains for their body size. A corvid has about the same brain-to-body ratio as a chimpanzee, even if the actual volume of its brain is only about the size of a walnut. These bird Einsteins lead complex social lives, with a strict social hierarchy, and even participate in social play.

Clayton's work has used these birds' propensity to hide their food for future consumption as a means to study their mental lives in many different ways. It's already well known that at least some corvids have an excellent memory, since they can recall the locations of hundreds of food items they have stored over previous months. But Clayton, along with her colleagues, has also shown that these birds plan for the future. For instance, scrub-jays, like us, enjoy a varied diet. If they know that their experimenter will only give them one kind of food the next morning, then the night before they will store a different kind in that location, thus ensuring, 12 hours later, that they have a choice for breakfast. Clayton has even demonstrated that they seem to have a clear mental understanding of the thoughts of competing birds. For instance, scrub jays will, if given the chance, watch where another bird has stored some food, and steal it away when the bird isn't looking. But if a scrub jay notices it is being watched by a potential bird thief, then it will furiously re-hide the food to fool the observer. You might think this is just an instinct. It isn't—a scrub jay will only attempt to re-hide food when observed if it has already learned how to thieve from other birds, and it will not re-hide it if observed by its mate. It's tempting to explain this attempt at disguising the food locations purely in terms of a sequence of clearly conscious thoughts that the jay could be having: First he discovers a wonderful extra source of food from exploiting all the hard work of another jay and stealing her stash; then he realizes that if he can steal a neighbor's bank of grubs, then so can any other jay; and finally, when he spots a potential thief, he imagines she could be harboring similarly thieving thoughts, and so he devises a strategy to confuse her.

Clayton has even shown that one species, the New Caledonian crow, can use tools in a surprisingly sophisticated way. For instance, these crows can use a sequence of tools in order to win food. In one example, the birds learn to use a short tool to hook out a longer one from a narrow tube, which they use to retrieve a third, even longer hook, which the crows finally use to directly fetch a food source. This was an experiment involving wild birds that voluntarily chose to enter the testing area. The crows do have a modicum of prior practice with the tools individually, but they have never before seen this sequential tool setup, and no human has ever demonstrated the task to them. Nevertheless, one of the three crows, known as Betty, successfully used the three tools in sequence and thus retrieved the food *on her very first trial.* Another of the birds, Pierre, also passed the test, but not in the way the experimenters ever intended: After exploring the apparatus, Pierre momentarily left the testing site, but he soon returned with a long twig that was perfectly suited to collecting the last, longest hook. This allowed him to retrieve the food using two tools instead of three. This form of complex, adaptive tool use suggested that the crows had a clear mental picture of the concepts involved and were invoking planning.

One final, particularly impressive example of tool use in corvids relates to the old Aesop's fable of the crow and the pitcher. The old story goes that a crow, almost dying of thirst, chances upon a pitcher full of water. But the crow can't access the water, because its beak can't get far down the opening at the top. After a while, it has a brain wave, and drops lots of pebbles into the pitcher so that the water level will rise and the crow can drink. This fable, it turns out, isn't fiction at all. This time, rooks were used instead of crows, but again no example demonstration by the experimenters occurred, and the birds were left to figure out the puzzle on their own. The rooks easily learned, usually on the very first trial, to drop stones into a beaker of water in order to raise the surface and reach a floating grub. To dispel any doubts that this was mere random behavior, the experimenters also showed that the rooks always stopped placing stones in the beaker once the food was retrieved; that the rooks favored larger, more effective stones over small ones; and that they avoided pointlessly placing stones into a beaker of sand instead of water.*

Not to be outdone by their feathered counterparts, an even trickier version of this ancient example of insight has been observed in the great apes by Daniel Hanus and colleagues. In this version, a peanut floats on a small volume of water toward the bottom of a beaker and can only be retrieved if more water is added to the beaker. There is a water dispenser nearby, which is about the only "tool" available. The cleverest 16 percent of the chimps realize the water's potential and discover an innovative solution by themselves, usually on the very first trial. They hold the water from the dispenser in their mouths and spit it out into the beaker the few times needed so that the peanut floats to the top and the tasty treat can be retrieved. One particularly inventive chimp, impatient to claim his food reward, decided to speed up this process by urinating directly into the beaker and winning the prize in this unusual way, seemingly unperturbed by its unpleasant liquid soaking, gobbling up the peanut as soon as it was available!

Intriguingly, in an earlier study the same research group found that every single one of the five orangutans tested could pass this challenge on the first trial, providing provisional evidence that this more distant cousin is in fact our smartest primate relative. Even more strikingly, four-year-old children are embarrassingly outcompeted on this task by chimpanzees and only show their superiority after reaching the age of six.

It seems almost certain that a highly active mental life replete with considerable planning and imagination is required to solve such devilish tasks. But does all this complex tool use mean that the crow family, effectively a primate in feathered form, is conscious? Are crows just as conscious as the great apes, who have similar mental skills? Based on their behavior so far, all we can definitively say is that corvids and the great apes are highly intelligent for nonhuman animals, while their seeming capacity for innovative thoughts is strongly suggestive of consciousness. But it isn't proof. We have no access to their mental world, and so simply can't tell for sure.

---

*For some videos of the crows using tools, see www.thenakedscientists.com/HTML/content/interviews/interview/1202/ and www.newscientist.com/article/dn17556.

CAN A BIRD ADMIRE ITSELF IN A MIRROR?

Another way to grapple with this problem is to set the consciousness bar challengingly high: It is widely assumed that if you are self-aware to the extent that you can recognize yourself in the mirror, then by definition you are also conscious. So, if an animal passes this demanding test, then there is little doubt it is conscious.

The standard way to test this is to put a colored spot on the animal's face in such a way that the animal can only detect it in a mirror. If the animal sees its reflection as itself, and not another animal, as many animals do, then it might try to rub the mark off its face, or at least move its body to get a better view of the mark. This is seen as positive evidence of self-awareness. The trouble is that an animal may fail the test for numerous reasons, regardless of whether or not the animal has self-awareness. It might have poor eyesight, might not be looking in the right way, or instead may well notice the mark, know that it is a spot on its own face, but not be inclined to investigate or remove it on that particular testing day. Humans don't pass this test until they are about eighteen to twenty-four months old, on average. The other animals that have passed it include chimpanzees, orangutans, gorillas, elephants, pigs (on a modified test), and even a member of the corvid family, the magpie. To my mind, passing this test really is strong evidence that an animal is conscious. But a determined skeptic would still have a valid position if he retorted that the animal was in no way reporting to us that it was conscious, and may merely have been reacting in a sophisticated way while inside there were no experiences whatsoever, no conscious thoughts or feelings at all.

But if we are looking for absolute incontrovertible proof, then we cannot even establish that any human, except for ourselves, is conscious! I think it's useful, therefore, to be somewhat pragmatic about the issue of animal consciousness. Although it's equally unhelpful to be absolutely convinced from scant evidence that other animals are aware, we should see hints such as passing the mirror-recognition test as strong, if problematic, evidence of animal consciousness.

What's missing is for an animal to report on its own level of awareness. But how can a nonhuman creature ever tell us that it is aware? In fact, some ingenious researchers have devised experiments where other animals can tell us exactly this—and given this chance, the animal even exhibits a level above pure awareness, to being aware *that it is aware* (in other words a kind of meta-awareness). So these animals demonstrate not just consciousness, but a sophisticated form of consciousness to boot.

For instance, recall that monkeys can be trained to tell you when their visual experience flips between one image and another on a binocular rivalry task. The stimulus is always the same, but the switch occurs within the mental life of the monkey, which it can report via a button press. The skeptic has a far greater challenge denying this form of awareness.

Gambling on Consciousness

However, there is even more impressive evidence of monkeys being able to report on their internal mental states. In a series of trials, Nate Kornell and colleagues showed monkeys a set of dots on a computer monitor, with one of the dots a little larger than the others. The monkeys would try to choose the dot that was the odd one out. They'd then be presented with a gambling choice of two buttons, one that was a safe option, and the other that was high risk, but with high potential rewards. If they pressed the high-risk button, three tokens would be added to a tally if they were previously correct on the dots task, but if they were wrong, three tokens would be removed. When this tally reached twelve tokens, the monkeys would get a special reward of a banana pellet. But the monkeys could also, after guessing the odd one out, choose the safe option, which always added one token to the tally, regardless of whether they were right or wrong.

A sensible policy here would be to choose the high-risk button when you were confident that you had guessed the dots task correctly, and the safe button if you really weren't sure. This sounds obvious, but grasping it requires a rather sophisticated internal mental world, where we have knowledge of the strength of our own beliefs, sometimes knowing our perceptual decisions are accurate, but at other times doubting the validity of those decisions. Monkeys, spurred on by the chance to obtain their tasty snacks, performed the gambling component of the task in appropriate ways, choosing the risky option far more when they were correct in the previous odd-one-out task, and the safe option when their perceptual guesses concerning the dots were less accurate. Not only this, but without further training on the gambling component, monkeys were easily able to transfer this ability for their confidence to track the accuracy of their perceptual decisions to other tasks, such as those involving working memory. They were also able to ask for hints when unsure, as we do, providing another strand of evidence in support of an inner mental world capable of having knowledge about your own knowledge level. For instance, when learning a complex sequence, initially the monkeys would ask for many hints, but when they became proficient at the sequence, they asked for very few.

In fact, the monkeys' ability to track the strength of their own knowledge, and use this skill to gamble appropriately (making risky bets when knowledge is firm, and safe bets when uncertain), was highly comparable to the average level of human skill on such tasks. So on this very advanced task, which in humans we take as an unequivocal sign of consciousness, monkeys and humans are equivalent.

Further bridges between monkeys and humans can be built by looking at what is occurring in the monkey brain during such gambling tasks. If these decisions in monkeys are indeed conscious ones, a reassuring result would be to detect neural activity that directly mirrors the differing confidence levels in regions we know relate closely to consciousness in humans. So far, this has only been studied in the back portion of the prefrontal parietal network, in the posterior parietal cortex, but in this region neuronal activity exactly matched level of confidence in a perceptual decision, just as if it were strength of awareness that were being measured.

Other species that have shown similar skills, where they gamble on high-risk options when they are usually correct and stick with safe options when they are less likely to be right, include the great apes and even, in a simpler version of the task, rats.

My reading of these results is that it is almost certain that monkeys, and any other type of animal that can also pass such tests, already have an extensive form of consciousness. But because monkeys are so similar to us in this gambling ability, we need to look elsewhere if we wish to explain exactly what distinguishes the immensely rich human form of consciousness from what is most likely a relatively weaker form in other species.

ANIMAL CHUNKING

So far, these behavioral studies have attempted to show that other animals are conscious in some form or other, but what these experiments have largely avoided is the question of the quality of consciousness—What can these animals actually do with their awareness? One interesting question along these lines, based on the main arguments in this book, is how well other animals can use chunking strategies. Humans are certainly not the only animals that can learn in a structured way. Rats, for instance, have been shown to apply simple forms of chunking. If you present a rat with 12 hidden openings, comprising 4 sets of 3 different types of food, they will learn to group the openings together according to the food group and head straight for the 4 openings that are sources of their favorite food. Even pigeons have been shown to apply a rudimentary form of chunking in their learning. For example, Herbert Terrace trained pigeons to peck either a sequence of colors or a mix of colors and plain white shapes. In order to obtain food, the pigeon had to get the entire sequence right. The pigeons were painstakingly trained for up to 120 sessions to get a sequence of 5 in a row correct in order to get food—probably the equivalent human task of completing a degree. The pigeons struggled terribly if the sequence was all 5 colors, but they did far better if they could break the sequence down into 2—for instance, if the first 3 were colors and the last 2 were shapes. It wasn't just that there were 2 different kinds of stimuli to remember: If the colors were interspersed with shapes—for instance, a shape, then a color, then a shape, then a color, and finally a shape again—then the pigeons were back to being terrible at the task. This grouping of parts of a sequence could well be analogous to human forms of chunking.

But there is a world of difference between being able to chunk at all, on the one hand, and being able to discover and use chunks in a powerful, hierarchical way, on the other. Other animals may be able to recognize themselves in a mirror, plan for future events, remember many past ones, or even be aware of their own awareness, but one key factor where humans—even toddlers—leave animals in the shade is the extent of our ability to chunk. Specifically, it seems that the number of levels of chunking on which we can operate, the height of our pyramid of meaning, easily outstrips even our closest relative, the chimpanzee.

One way of demonstrating this is to observe different species at play. In one experiment, a group of chimps from 15 months to adulthood, one adolescent bonobo, and human babies between 6 months and 2 years of age were all given a random collection of 6 objects, such as cups, rings, and sticks that could be red, blue, or yellow. The experimenters simply watched them play and recorded how they moved the objects about, combined them, and so on. Various levels of information and behavior are available here: On a basic level, these are all separate objects and can be manipulated one at a time. But far more meaning than this can potentially be extracted. For instance, the objects can be grouped by type, size, or color alone, but two categories can also be combined—for instance, by placing all the red rings together. A higher level still is even available, if all the large red rings, say, are separated from their smaller equivalent. Then there are the relations between different items. For instance, sticks and rings can be placed *inside* cups, and sticks can pass *through* rings.

When it comes to the basic skills of manipulating single objects, there is little to separate human babies from their primate cousins of the same age. As soon as you move up the information pyramid, though, developmental differences become increasingly apparent. Chimps and bonobos can learn to group items according to categories, such as color or shape, but they learn such concepts considerably later than humans do, and the complexity, or levels, of meaning they can learn are terribly limited compared to human babies. It is not uncommon, for instance, for a human baby, by the age of two, to use one hand to turn upright and then hold a cup, and use her second hand to clasp a few small spoons together out of a set of random cutlery, which includes larger spoons, forks, and so on, before finally placing the small spoons in the cup. By grouping the small spoons together, she is demonstrating at least two conceptual levels above individual items because she is picking objects based on two combined categorical features. She is then demonstrating knowledge of another level again when she puts the spoons in the cup, by linking a group with another item. These seemingly simple acts appear largely beyond any other primate, at any stage of development. But, of course, humans rapidly learn far more complex, hierarchical concepts and actions than this relatively straightforward example.

In a more formal series of experiments exploring this issue, humans between the ages of 11 and 36 months and mature chimpanzees, bonobos, and capuchin monkeys were all compared on the same simple task. All subjects were given three nesting cups, a standard children's toy (see Figure 7). The experimenter repeatedly demonstrated what was required of the subjects: To place the smallest cup inside the middle cup, and then put both cups together inside the largest one. The cups were then dismantled and the subjects were encouraged to copy the experimenter exactly. This is a useful developmental test for human babies. When they are around a year old, all that most children can do is place one of the three cups inside another, leaving the third untouched—in other words, they can't complete the task. By around 16 months, they can complete the task, putting the middle cup inside the large one and then the smaller one inside the other two. This isn't quite what the experimenter showed them. Critically, these toddlers haven't yet grasped the complex, hierarchical idea that you can move two cups at once (the middle one with the smaller one inside), as if they were a single compound item, and in addition that this group of cups can still be placed in any cup larger than the outside one. (I've watched my daughter at this age play with the same kind of toy, and even if I place the smallest cup inside the middle one to help her to solve the puzzle, she will deliberately take the smallest cup out before stacking the cups one by one [middle into largest and then smallest into middle]. For her, it seems as if the concept of groups of nesting cups is seemingly impossible. The group simply has to be dismantled for progress to be made.) Finally, many children from around the age of 20 months or so can exactly copy the experimenter. This shows that they have grasped the idea of hierarchies, so that two objects put together can in some sense be seen as one single object.* You might suspect that part of their mastery of this progression arises simply from infants and toddlers having ample time outside of the lab to know how

---

*The original scientist who ran this study, Patricia Greenfield, provided evidence for the thesis that this staged development of abilities to chunk movements mirrored a child's ability to learn the complexities of language—for instance, combining subcomponents of words together to form more complex words, and using multiple words in a grammatical structure. This task therefore is another clue that our ability for language might just boil down to our ability to chunk, especially in a hierarchical way.

to manipulate these or similar items. But although that's true of most babies in Western culture, this and a related experiment were also carried out in Zinacantecos babies and toddlers in southern Mexico, and they also exhibited mastery of hierarchy. This Mayan group has few materials in its environment on which to practice these kinds of manipulations, and the children have no toys, but the Zinacantecos children showed exactly the same pattern of development, suggesting that this stepwise acquisition of the mental machinery of hierarchical chunking is universal.

What happens when the chimps, bonobos, and monkeys try this task? Much of the time, even with extra, guided training, they only reach the first developmental stage that the children reach. A minority of the bonobos can at least reach the intermediate, nonhierarchical stage of stacking all three cups correctly, like the human sixteen-month-olds. But virtually none of the animals can grasp the idea of combining two objects together and then moving them, even though the experimenter had just that moment shown them exactly this.

Therefore, young children can outperform adult chimps, bonobos, and monkeys on tasks requiring that they chunk in this hierarchical way. Although these other animals may well be highly conscious, the exceptional richness of human awareness may critically be reflected in our superior ability to find and combine structured information.

Of course, stacking cups, while surprisingly complex, is nevertheless one of the simplest tasks human children perform. As children grow up, the toys they play with rely on an increasing number of levels of meaning as well as more sophisticated relationships between items. And each stage of human play reflects the widening gap in cognition and consciousness between children and primates of the same age.

I described in Chapter 5 how there is nothing exceptional in the quantity of items humans can place in consciousness: Humans, like a wide variety of other species, are limited to only three or four working memory objects. However, human consciousness is so rich, so powerful, because of the extent to which we can manipulate and combine this handful of online items, especially in hierarchical ways. The above experiments demonstrate how this seemingly trivial difference between humans and other species starts conceptually to explode after only a couple of years of life.

INFANT AWARENESS

Does this mean that human infants aren't conscious until they are around twenty months old and are able hierarchically to combine objects and actions together? Almost certainly not. This milestone merely signifies that a critical stage in a growing consciousness has been reached—a stage where experiences will be far more varied and complex, and where learning can skyrocket.

When, then, do the first seeds of infant awareness start reaching up from the soft soil? Does this occur before birth? A fetus can be remarkably active, kicking and punching away on a regular basis, from surprisingly early in pregnancy—usually by about seven or eight weeks, although the mother won't feel these movements until a few months later, when the fetus has sufficiently grown. Toward the end of pregnancy, the fetus can also be highly responsive to the outside world, either via pressure on the uterus or muffled sounds filtering in from outside. Does all this signify consciousness? This is unlikely, because both the mother's placenta and the fetus itself work actively and in concert to keep the fetus under safe sedation while inside the uterus. Effectively, the fetus lives its prebirth existence in a kind of dream state—though with few stimuli so far absorbed, such dreams would be unlike ours, with little, if any, detail. Instead, the fetus only really wakes up in the sudden, shocking moment of birth.

Is this the moment of first consciousness? This is where the behavioral approach fails us, as there are few clues early on that clearly demonstrate awareness. My personal intuition, from watching my own baby daughter develop, is that you can't help assuming that there is a strong sense of awareness from soon after birth. If consciousness cares about novelty and unexpected events, then my daughter's profound surprise for much of her first two months whenever she developed the strange sensation of hiccups is one intriguing piece of evidence in support of early consciousness. Then, just shy of her third month, she started laughing at my silly antics. Admittedly, with little science to back this up, I felt that this humorous reaction was the signature sound of conscious surprise, and it gave me little doubt that my daughter was indeed now aware of the world.

Still, another approach that ignores behavior entirely might provide more definitive answers to such questions where they otherwise cannot be found.

MEASURING CONSCIOUSNESS IN ANIMAL BRAINS

Assessing consciousness in other beings via these roundabout behavioral measures is a fascinating thing to do, but there are problems with this approach. The first, which I have already mentioned, is that it is difficult to interpret a positive result. Because the animal cannot tell you that it is conscious, any pass in a test has to be taken with caution as to its implications for the exact nature of the animal's inner mental world. This is more readily true in artificial intelligence, where a robot could be programmed to pass any of the tests I mentioned above, from the mirror-recognition test to the gambling tasks. This signifies very little. The robot would probably fail any other task of awareness, passing only the one it was programmed for, and of course a few lines of code does not equal consciousness.

The second problem is that we cannot rely on a negative result in these tests either. In the novel *The Girl with the Dragon Tattoo* by Stieg Larsson, the title character, Lisbeth Salander, is quite a wild child. The teachers and authorities investigate her unruly behavior by subjecting her to every psychological and educational test in the book, but her response is to refuse even to lift up her pencil to write her name. Because she effectively fails each of these tests, the authorities assume with surprising dogmatism that she must therefore be mentally retarded, and she is consequently officially classed as such well into adulthood. The truth, it increasingly transpires, is that she is in fact fiercely intelligent, and she is quite capable, when the need arises, of twisting these prominent authority figures around her highly independent finger. What the authorities should have understood when testing this girl was that if someone fails a test, an inability to carry out the required attribute is only one of a range of possible explanations.

Similarly, if an animal is not interested in playing along, you have no way of knowing if it is capable of passing the test. Maybe it could easily pass it, but stubbornly refuses to try. In other words, as with so many tests in psychology, a negative result simply cannot be interpreted as an inability of the animal to perform a given skill. Another problem with the behavioral approach is that there is no clear way to gauge *how* conscious an animal is. Only a few very distinct stages are explored, and there is little scope for a continuum of conscious level, which is probably a more appealing idea than the more ugly assumption that you either have consciousness or you don't.

There is a way to sidestep these traps, and that is by ignoring behavior and merely investigating the structure and function of a brain, or indeed any computational object, for telltale signs of awareness. The first, crudest attempt is simply to rate an animal according to its brain size. If we use such an index as a rough estimate of consciousness—perhaps by virtue of the provisional argument that more neurons means that more information can be processed in a given brain—then humans come near the top of the table, but are by no means the leaders of the pack. Our brains weigh about 1.3 kilograms, a shade less than the bottlenose dolphin, whose brain weighs in at 1.8 kilograms. The African elephant has a brain that weighs around 6.5 kilograms, nearly five times that of the human. Of all the animals on land or sea, the sperm whale wins comfortably, with a brain that tops 8 kilograms. So if it's true that brain size alone reflects consciousness, then an intriguing thought is that dolphins, elephants, and some whales have considerably more of it than us.

But in some ways it's not surprising that a sperm whale has a brain six times the size of our own. After all, sperm whales can weigh nearly a thousand times what we do. Much of that extra brain mass is probably needed to move a body that's 20 meters long, as well as to keep track of all the other internal states that need to be managed rather more carefully when an animal is larger than most buses. Because of this, most scientists believe that a better comparison to make is the size of a brain *compared to the animal's body*. The logic behind this is that if a brain is far larger than you'd expect from the animal's body, then all those extra neurons must be doing something over and above the standard tasks of making the animal move, regulating its states, and so on—and very probably the extra brain matter relates to more complex processing, including consciousness.

Although calculating this brain-to-body ratio is rather more complex than it at first sounds,* humans come out on top in the entire animal kingdom, with

---

*It turns out that there are many factors that need to be taken into account for an accurate, useful ratio of brain to body to be determined. For instance, recent evidence suggests that some species are far better at packing in more neurons per brain weight than others, so measuring brain weight alone isn't very informative.

a considerably larger brain than you'd expect for our size and body type. Dolphins aren't too far behind, followed by chimps, bonobos, orangutans, and gorillas. It's broadly assumed that an animal's position on this line is a reasonable reflection of its ability to learn, but we only have circumstantial evidence that our large brains really do reflect our greater consciousness. We also have no idea how consciousness scales with brains—the threshold for consciousness may be a hair's breadth under the human brain-to-body ratio, or a billionth of it—if even such a threshold exists. Therefore, this biological comparison can only be seen as a suggestive hint of how conscious an animal is.

CHAUVINISTIC ANATOMICAL BOOTSTRAPPING

A more promising approach is a kind of bootstrap method, where we learn from humans what parts and processes of the brain are important for consciousness and we then assess the level of similarity between these key features in humans and other animals. The thalamus and prefrontal parietal network are crucial for consciousness in humans. So to what extent do other animals share these structures with us? All vertebrates have a thalamus in some form, but not all have any kind of cortex resembling the prefrontal parietal network. From this line of evidence alone, we'd conclude that our great ape cousins, with a prefrontal parietal network not dissimilar to the human model, have the most similar levels of consciousness to us. Other primates, such as monkeys, have prefrontal and parietal structures that we can broadly match with our own, but the anatomical stretch suggests that their conscious capacity is diminished compared to ours. Most mammals at least have a cortex—and some capacity for consciousness, perhaps—while nonmammals, with little hint of a prefrontal parietal network, may not have any consciousness at all.

But this approach, while at least adding more clues to the collection, feels rather circumstantial. It also discounts the possibility that consciousness could arise in animals with a very different brain to ours.

We tend to think of all those animals that never left the oceans as far more mentally simplistic than us, almost certainly not conscious—which is part of the reason why many people still consider themselves vegetarian if they eat fish. But the octopus's cognitive skills, if fully known, would raise doubts in many who believe such assumptions. The octopus, although an invertebrate—with no thalamus or cortex to speak of—behaves in ways that utterly belie its primitive label. It has around 500 million neurons, not too far from the numbers in a cat. But the octopus brain is decidedly unusual, with an exceptionally parallel architecture—almost always a positive quality when you are talking about brains. The majority of octopus neurons are to be found not in its brain, but in its arms. In effect, if you include the neuronal bundles in its limbs, the octopus has nine semi-independent brains, making it unique in the animal kingdom. The octopus is also a genius among ocean creatures. It has highly developed memory and attentional systems. In nature, this allows these invertebrates to take on a wide range of shapes to mimic other animals, rocks, or even plants. In the lab, octopuses can distinguish shapes and colors, navigate through a maze, open a jar with a screw-on lid, and even learn by observing the behavior of another octopus—an ability thought previously only to exist in highly social animals.

David Edelman, who studies octopus cognition with Graziano Fiorito, has spoken of his uncanny experiences upon entering the octopus room in the grand pillared basement of their palatial Naples zoological department. All of the octopuses immediately press their faces to the sides of their tanks and carefully, continuously track the movements of this new intruder. Such sustained attention is normally only found in obviously intelligent animals. If octopuses are conscious of their world, then we would simply never realize it from this comparison of brain anatomy, as their brains are utterly unlike human or even mammalian brains.

QUANTIFYING CONSCIOUSNESS

While all these approaches in concert help inform the debate about animal consciousness, one theory stands out in its promise of a definitive solution. Giulio Tononi's information integration theory is a well-regarded, modern theory of consciousness. It promises a single number for conscious level, calculated from the measure of the number of neurons in a brain, how they are wired together, and how they interact. In principle, this could show, to pluck numbers out of the air, that fully awake humans have a consciousness of 100 units, coma patients 2 units, chimps 50, rats 10, and so on.

A clear consequence of this theory is that virtually every animal will have some value for its consciousness level. A honeybee, for instance, with nearly a million neurons, certainly will. Even the simple nematode worm, *C. elegans*, with its 302 neurons, will have a value for its level of consciousness, although this figure will admittedly be minuscule. It's even conceivable that a colony of ants would, under this system, be collectively classed as conscious. Some people are uncomfortable with the notion that such lowly creatures could even have a minimal level of consciousness, let alone a group of animals. Although it is true that more work needs to be carried out to validate this theory, it may turn out that such skeptical intuitions may be wrong. It may well be that any kind of brain, however small or simple, will generate some level of consciousness. The fruit fly, for instance, shows signs of a rudimentary attentional system, which is certainly one of the prime mental components of consciousness.

Tononi's information integration theory is also compatible with the notion that computers or robots will at some point have consciousness—and we could in principle use the mathematics of the model to rate the level of consciousness of some artificial being according to its network equivalent of a brain.

However, this theory—and indeed all current theories linking consciousness with joined-up information in a network—rules out consciousness in bacteria and plants, despite their rudimentary computational processes. There simply is no information network to speak of, nor, returning to my main thesis, is there any capacity for the creature, in the moment, to combine lower-level information to form a more meaningful chunk.

In practice, things aren't so simple. As the theory stands in its current form, the number of computations required to calculate this consciousness number scales up ferociously with the number of nodes, or neurons, so that even with a simplistic simulation of the humble *C. elegans*, with its 302 neurons, it would take 5×10^{79} years on a standard PC to calculate its level of consciousness—by which point the universe may no longer exist! Whatever the other strengths of this theory, unfortunately it cannot practically be used to measure the conscious levels of any animal, and certainly not of humans.

However, there are researchers, including my current lab colleagues Adam Barrett and Anil Seth, who are working hard to adapt the theory to make it practically computable for large-scale systems like the human brain. So in all likelihood there may be effective ways of calculating the level of consciousness of any being within the next decade, at least based on Tononi's theory. All that would be required would be a mathematical algorithm applied to an approximation of the number of neurons (or artificial nodes) and the connections between them in a given brain—data that are already available for many species.

Rather than waiting for the mathematicians to adapt this measure of consciousness so that the calculations don't take the age of the universe, Giulio Tononi, along with his colleague Marcello Massimini, has already been developing an intriguing method by which to make a practical rough-and-ready approximation. The experiment uses EEG to record brain waves as well as a transcranial magnetic stimulation (TMS) machine. TMS involves a figure-eight device about the size of a small book that is placed on the scalp. This machine is in fact a powerful electromagnet. The magnet is turned on for a fraction of a second, which causes the cortical neurons under the scalp, at the center of the figure-of-eight coil, to fire. All that the subject feels (I've experienced TMS myself multiple times) is a kind of tap on the head. In this particular experiment, the only task the volunteers had to perform was to nod off while the TMS pulses were delivered every couple of seconds. If the subjects were awake, the TMS would cause a spike of brain waves that would spread almost all over the brain during the next few hundred milliseconds or so. When the subjects drifted off to a dreamless sleep, although the initial spike of activity was greater, it would die down faster, and would remain only at the local site that was stimulated by the TMS machine. This study shows firsthand how, in wakefulness, information can flow freely across our entire cortical surface, but when we're asleep, although the neurons are just as capable of firing, they only weakly transmit their information to their nearest neighbors. Massimini and Tononi see this as evidence that our ability to combine information throughout much of the cortex is high when we're awake, but low when asleep (these data are also applicable to other consciousness theories, however).

Although at present this method can do little more than distinguish between wakefulness and sleep, plans are afoot to turn it into a far more sensitive measure. Massimini and his group are starting to use the complexity in the EEG wave, rather than how far it travels and for how long, as a more accurate way of gauging level of consciousness. Soon one of these techniques may be able to generate a practical index of consciousness. This quantification of the level of awareness could be applied to any normal subject, whether awake or asleep, as well as any patient, or indeed many animals. Similar methods could even be developed in the future to perturb the activations of an artificial being. Therefore, within the next decade or so, we may well have a viable means of measuring and comparing consciousness in humans and many other animals.

ETHICAL IMPLICATIONS

How can a science of consciousness assist the moral conundrums surrounding issues such as animal rights, or human abortion? This is certainly not a book on ethics, and these questions are fiendishly complex. All I will offer here, then, are a few personal thoughts on the matter, informed by my understanding of the science of consciousness.

In ethics, a broad distinction is often made between two very different frameworks: The first is a rights-based system, with laws against murder and theft being obvious examples. The second type of ethical mechanism, in contrast, centers only on a calculation of the net pleasure and suffering of a given population. Economics and its obsession with money would be a rough approximation to this second stance.

If we were to talk about the first system, of rights, such as the right to life and to freedom from easily avoidable suffering, then any animal that has a broad potential for consciousness would also have a significant capacity to suffer, and should fall under the umbrella of such rights, preferably as enshrined in law. Personally, I would want to live in a society that would err on the side of caution in order to ensure that suffering in innocent creatures by our powerful hands was minimized.

One international movement, the Great Ape Project, has aims that mirror these, though with limited scope. Backed by renowned scientists such as Jane Goodall and Richard Dawkins, this movement is pushing for a United Nations declaration to ensure that all great apes (chimpanzees, bonobos, orangutans, and gorillas) have a right to life and freedom from torture. I believe that based on the current scientific picture of animal consciousness, governments around the world should not only accept this view but also seriously consider extending its scope. In the scientific thesis that I have defended here, consciousness is most closely aligned with innovation. Tool invention and use, which require innovative, flexible thoughts, are therefore strong indicators of an extensive consciousness. This would class not just our great ape cousins, but also, at the very least, monkeys, corvids, dolphins, and octopuses as creatures that deserve protection under our laws. Experiments have additionally shown that various nonhuman species can master self-recognition in a mirror and demonstrate self-doubt. Given that we'd automatically take these skills as evidence of a rich form of consciousness in ourselves, we should cautiously accept the same conclusion for any other animals that use such abilities. The list of animals that use tools, recognize themselves in a mirror, or exhibit self-doubt would currently include not only the great apes, but also dolphins, monkeys, elephants, pigs, corvids, and octopuses—although the list will almost certainly grow as more tests are carefully carried out. Barring all these animals from being subject to experiments that would cause suffering, removing them from our food industry, and making it a crime to harm or kill such animals would be a radical step, and not one that I can see any political leader advocating any time soon. Nevertheless, it would be a consistent and caring departure from the way we currently view animals, and would acknowledge the advances in our scientific understanding of the mental lives of these other species.

What of the second form of ethical system, where any discussion of rights is ignored, and all that matters is the net pleasure and suffering in a population? To illustrate the details of such an ethical framework, imagine that a million mice each consciously suffered amount x in the lab as part of research that could lead to the eradication of a human disease. This sacrifice would be justified if, say, 100 humans are each spared at least 10,000 amounts of x in conscious pain, or if at least a million people would no longer each have x amount of suffering (in both cases, at least a million amounts of x suffering is avoided).

Under this ethical system, we would clearly need, somehow, to be able to quantify the level of consciousness, and therefore the amount of suffering, that each animal actually experienced. Although we are inching toward the ability to calculate conscious levels using current theories and techniques, we are certainly not yet at this scientific stage. However, we should keep our eyes wide open for advances in the next decade or so. Once any practical means of making these calculations emerges, we shouldn't hesitate to apply these techniques to any part of our culture where animal suffering may be occurring. This way, we would ensure that we were not unduly creating net amounts of suffering in an extended population of all conscious creatures, and that we were guarding against underestimating the conscious levels of other animals while selfishly exaggerating our own capacity for suffering.

Wider ethical issues—for instance, involving abortion and the right to life, as well as artificial consciousnesses—could also be evaluated on the basis of some quantification method for level of consciousness. But it's worth bearing in mind that for fetuses, despite our intuitions, all the evidence currently suggests that consciousness is highly unlikely before birth. As for artificial intelligence, no computer or robot in the world currently has an architecture that could even provide the smallest form of awareness, let alone allow for suffering, which might even be an entirely distinct issue in artificial life. It might be possible, for instance, to engineer an artificial being that is conscious and motivated, but that lacks the capacity for anything beyond the mildest levels of suffering, because we haven't programmed it for anything more intense.

DIFFERING QUALITIES AND QUANTITIES OF EXPERIENCE

The two parallel probing streams of this chapter, one behavioral and the other physiological, paint a broadly convergent picture of animal consciousness, though with some marked differences. The behavioral approach, ignoring any hints from brain size, neural complexity, and so on, provides tantalizing provisional clues to consciousness in very many species, possibly even down to the lowly fruit fly with its rudimentary attentional system. But far firmer evidence of awareness, such as when animals recognize themselves in the mirror, or can report on their own level of doubt, is limited only to that handful of species with the largest, most sophisticated brains. It seems reasonably clear from such behaviors that these privileged species can demonstrate an advanced form of consciousness. A little surprisingly, from our usually chauvinistic perspective, that set of species that we would class from their behavior as almost certainly conscious isn't limited to our closest evolutionary cousins, the great apes, or even a little further afield, to primates and other higher mammals.

The physiological approach sidesteps problems of reliability inherent in making observations of behavior, but it also creates new problems because physiology is a far less direct measure of consciousness than behavior. There is also less scientific consensus about which of these indirect physiological approaches is the most accurate. But perhaps the most promising direction would be something along the lines of information integration theory, where the complexity of a system, regardless of how anatomically similar it is to humans, would be an index of levels of consciousness. On this basis, there would be a continuum of consciousness, with humans, and our most densely interconnected brains, at the top, but some of the simplest animals having some level of awareness as well, albeit a minimal one.

Both the behavioral and physiological approaches suggest that those animals with the largest, most complex neural architecture and the most able minds are the most conscious. But each technique emphasizes a different perspective on the question of nonhuman awareness. The most parsimonious way to marry the two methods is to conclude that there is indeed a continuum of consciousness from humans to the smallest, simplest of creatures, but that within this there are distinct, meaningful steps that emerge out of a deeper sense of awareness. These steps include self-awareness, a sense of doubt, and the ability to build pyramids of meaning within our own minds.

This final skill, unique to humans, may underpin our flair for language. It is reflective of our exceptionally flexible working memory, and is very probably the underlying reason for this nagging sense many of us have that humans really do have a clearly raised awareness compared to any other species.

This explanation for what crucially distinguishes humanity from the rest of the animal kingdom fits neatly with the technological marvels of the modern world that pervade our lives. These advances are a constant reminder of how different we are from other animals. Such a view of our unique qualities could partially explain human evolution. Early precursors to *Homo sapiens* may well have used their fledgling ability to spot patterns to powerful effect, creating primitive technologies that would have been impressive to the tribe and yield a greater variety and quantity of food. A technological and evolutionary runaway effect could easily have occurred, giving rise to the modern human brain, which has an immense, diverse consciousness and a profound, unique ability to chunk information, to notice and exploit deep patterns in nature, and to create all the technological marvels that enrich our lives and make it so much easier to survive than ever before.

But the price we pay for such a unique, expansive awareness, supported by a supercharged neural machine, is intense fragility. Our large, immensely complex brains can be easily, irreparably damaged, sometimes by just a knock. Such accidents can diminish or even permanently rob us of consciousness. Another form of neural frailty is far more subtle, but also far more prevalent: Genetic or chemical abnormalities in our brains can easily upset our mental balance and breed any one of a range of insidious mental illnesses. In the final two chapters I will describe the often tragic corollary of our collective capacity for deep conscious innovation. I'll first detail precisely how vulnerable our brains are to significant damage and how current ideas of awareness can help detect what consciousness remains following traumatic brain injury. Then I'll outline how a dysfunctional consciousness lies at the heart of most forms of mental illness, and close with how the model of awareness espoused in this book can help illuminate a path out of these emotional prisons.

7

Living on the Fragile Edge of Awareness

Profound Brain Damage and Disorders
of Consciousness

JUST TOO COMPLEX?

When recruiting normal volunteers for our neuroimaging experiments, I always ask my potential subjects a series of questions to make sure they are suitable. For instance, because the fMRI scanner is essentially a tremendously powerful magnet, if a volunteer has a metal implant, then even approaching the scanner could be dangerous. We also ask questions designed to ensure that each person in our eventual group of participants has as normal a brain as possible. For instance, we exclude anyone who has had brain tumors, strokes, or any other neurological condition.

But in addition, we rule out anyone who has ever been knocked out, even if it was just for a few minutes. This excludes a large proportion of people, as you'd imagine, since head trauma is a common side effect of participation in sports, for example. But this step is necessary, because the sad fact is that a single concussion can easily cause low-level brain damage. And in about 10 to 15 percent of cases, especially if any later concussions occur, there will be more severe brain damage, causing long-term or even permanent memory and concentration problems. Recent evidence suggests that early Alzheimer's disease can even be induced.

The huge human brain is particularly susceptible to trauma. Brains have the consistency of jelly, and if we suffer a violent blow to the head, our brains bounce back and forth inside our skulls, twisting and shearing in the process. The wires connecting our neurons stretch and strain, and they are easily damaged. The outer edges of the brain, especially at the front, can scrape against the sharp parts of the inner skull and be destroyed. Almost all other species are far less likely to suffer concussions than humans, because their brains are so much smaller and are not as susceptible to such bouncing, tearing forces.

In addition to our vastly increased risk of concussion, we face the risk of more severe brain damage from a serious blow to the head (or a host of other causes). Impacts can all too often lead to a vegetative state, coma, or even death. Such patients' brains can look decidedly warped and thinned, with little cortical matter left. Entire lobes can be missing following a fall or car accident. Just looking at the MRI scan, though, cannot prepare you for meeting the patient in person. Severely brain-damaged patients in a vegetative state are normally wired up to a large set of medical machines. Their limbs can be twisted, with the body twitching and writhing intermittently. Unnatural, repetitive movements are common. Although they may have their eyes wide open, and superficially seem awake, a close inspection shows that this is a cruel illusion. There is no real evidence that there is any kind of awareness of the outside world: The patients' eyes do not seem to focus on important features of their surroundings; their expressions are not responsive to the environment; and there is a general, frightening void when you try to infer their thoughts from any of their random actions. It is a heartbreaking and disturbing sight, and almost more difficult for the family than if the patient were in a coma. With the vegetative state, the outward signs of wakefulness constantly promise a hidden consciousness inside patients' heads, or at least hope of recovery. But in many cases that hope is false.

Now, knowing what I do about the brain, and having seen such patients, action movies are no longer just the mindless, cartoonish entertainment they were to me as a child: Whenever I watch an Indiana Jones or James Bond film, where countless nameless guards are victims of knockout blows by fists, wrenches, frying pans, and so on, I can't help wincing a little at each blow. To me, now, these attacks mean millions of neuronal wires being stretched to breaking point.

I'm firmly of the belief that scientific research is worthwhile, purely for the knowledge that we gain about ourselves and the universe, even if that fresh understanding can bear absolutely no practical fruit. But for a question as intimate and profound as "What is consciousness?" it's almost inevitable that society will benefit from the answer in a multitude of ways.

One of the subfields currently bearing fruit is the research in disorders of consciousness, with vegetative state one of the most prominent examples. Twenty years ago, a patient appearing to be in a vegetative state would have had surprisingly meager help from the medical community, beyond being kept alive. An assessment of just how conscious the patient was amounted to little more than guesswork. The same was true of any prognosis for recovery. Now, for diagnosis and prognosis, at least, the territory looks very different. A large, active research community is working hard to use increasingly mature models of consciousness to quantify the degree to which awareness has been stolen from these patients. Sophisticated predictions of recovery are also emerging. And although treatments are somewhat lagging behind, nevertheless there is tentative progress in this realm as well. In this chapter I will relate the advances that have been made in understanding and treating such severe disorders of consciousness. I will also describe how vegetative state research is reaffirming and extending our current ideas about how awareness functions.

A Tortuous Battle of Uncertainty

Terri Schiavo was a shy girl living in Florida. By her mid-twenties, with a reasonably uneventful, settled life, she and her husband, Michael, were trying to have children. However, Terri was probably suffering from severe bulimia as she fought to control the weight problems that had been an issue for much of her life. On February 25, 1990, in the middle of the night, Terri's heart stopped, most likely due to a chemical imbalance caused by the bulimia. The ambulance crew managed to revive her, but too many minutes had ticked away in temporary death, and she suffered massive brain damage.

Terri was in a coma for about three months. Her condition then improved slightly and she entered a vegetative state, where she could open her eyes and show signs of sometimes sleeping, sometimes waking. But this glimmer of hope was painstakingly eroded as it gradually became clear that she would not improve any further.

Eight years after the accident, with no change in her condition, and any remote hope of recovery having long since receded, her husband, Michael, filed a petition for Terri's feeding tube to be removed so that she could pass away. Terri's parents, in total contrast to the doctors' assessment, believed that Terri was still in there somewhere, that she could at times find pleasure in the company of her family, and that she might well recover further with some future treatment. Thus began one of the most famously bitter legal disputes in history.

Over the next seven years, there were numerous legal petitions, appeals, motions, and so on, with Michael, on the one hand, trying to let Terri die, and Terri's parents, on the other, determined to stop him. The case violently pitted civil liberties against religious values. Terri was a devout Catholic, and questions were raised as to whether she would have wanted her life to be ended in a potentially sacrilegious way. The pope himself ruled on the case at one point, stating that Terri should be kept alive. At times the story looked like a sorry soap opera, with claims being bandied about that money was a driving motivation on both sides of the family divide—Terri's estate was worth a considerable amount by this stage, because of successful litigation against the fertility doctor who had failed to spot her bulimia. At the same time, since 1993 Michael had been dating another woman, with whom he'd had two children, one in 2002 and the second in 2004, but he refused to get a divorce from Terri. His stated reason was a continued emotional attachment to Terri, but her parents claimed this was simply a strategy to lay claim to her money.

As the years rolled by, the feeding tube was removed numerous times, only to be replaced on Terri's parents' appeal. The public became increasingly fascinated by the case, and the politicians began to step in, realizing that cheap votes from the far right could be won by taking a firm pro-life stance. Jeb Bush, the governor of Florida, passed legislation giving him the right to intervene in the case and prevent Terri's death, although this was later found to be unconstitutional. Eventually even the president, Jeb's brother, George W. Bush, weighed in on the pro-life side, canceling a vacation and flying back to Washington, finally signing emergency legislation late in the night, in his pajamas. The main purpose of this bill was to keep one person alive: Terri Schiavo.

In the end, the political and legal wrangling fell irreversibly on Michael's side. Terri Schiavo died on March 31, 2005, thirteen days after her feeding tube was removed for the last time.

At the heart of this long, painfully public battle were two heart-rending questions. First, was Terri Schiavo still there, even in some shrunken form, as the personality by which she had been defined prior to her collapse? And second, if the brain damage had indeed stripped her of any chance of normal awareness, might there have been a treatment on the horizon that could have restored it? Terri Schiavo is by no means a unique case—there are in all likelihood well over 100,000 similar examples worldwide, with the families of each of these patients obsessively asking themselves these two exact questions.

A Thin Veil Between Life and Death

The first stage in assessing how little consciousness is left in patients like Terri is by arriving at a formal diagnosis. If a patient is entirely unconscious with eyes persistently closed, then he is in a coma. If he has some form of sleep-wake cycle, and sometimes opens his eyes, but shows no signs of awareness, then he is in a vegetative state. Many vegetative state patients can make at least a partial recovery and even regain normal consciousness, but about half do not, and the longer the patient is in a vegetative state, the less likely he is to recover. In line with research on how normal consciousness is supported in the brain, the vegetative state is closely linked with damage to the thalamus, along with its connective pathways to the prefrontal cortex, especially when there are no signs of long-term recovery. But this syndrome is also linked with many other damaged brain areas, along with a range of initial causes, including head injuries, infection, drug overdose, and so on. If the vegetative state continues for a month, then it is classed as a persistent vegetative state. If it continues for a year (depending on the injuries and the country of diagnosis), then the patient is classed as being in a permanent vegetative state, or PVS. At this stage, chances of recovery are increasingly slim.

But if the patient's condition does improve and he shows some slight signs of awareness, such as tracking an object with his eyes, or responding to commands to move his arm, then he is upgraded to being in a minimally conscious state. It's relatively straightforward to ascertain whether a patient has moved from coma to vegetative state—you just have to see whether his eyes open and close—but it's notoriously difficult to tell the difference between a vegetative state and a minimally conscious state. Although the vegetative state might appear to be an improvement over a coma, the truth is that consciousness may still be entirely absent. Therefore, to provide reassurance to the patient's family that he is consciously present and might recover, an improved diagnosis of minimally conscious state is crucial.

For the patient's family, it's easy to mistake certain outward signs as hopeful evidence of consciousness—the patient might weep, smile, or grind his teeth—but in fact these are effectively reflexes and can occur with no consciousness whatsoever. Doctors, too, although mindful of overinterpreting these surprisingly primitive behaviors, can often struggle to disentangle the subtle, inconsistent signs of real awareness from random sounds and movements. Clinicians distinguish between PVS and minimally conscious state by looking for any behavioral hints that the patient is conscious. Can he speak at all, for example? Does he seem to follow your eyes? Does he make purposeful movements? Their search for such evidence is within the usual challenging context of a patient who produces copious twitches or spastic limb movements and who might be drifting in and out of sleep on a regular basis. This somewhat haphazard approach means that it is extremely difficult to diagnose such patients accurately—with some estimates suggesting that 43 percent are misdiagnosed by the standard clinical bedside observations, usually because patients are assumed to be in a vegetative state when they should be upgraded to minimally conscious. Better behavioral assessments are slowly filtering through in the form of carefully designed rating scales, which standardize and quantify the diagnostic markers that place a patient in one group or another. But even this improved measure can be completely misleading in some cases.

VIEWING CONSCIOUSNESS FROM WITHIN

When we watch a friend open her mouth wide, raise her eyebrows as high as they can go, and point an urgent finger at a naked man wearing only a top hat who is going down the middle of a busy street on a unicycle, we have a

good idea as to the contents of our friend's mind, and have no trouble knowing that she is awake and normally conscious. But it's easy to forget that this
is only ever an indirect measure: Her hand movements or the expression on
her face are merely the *outputs* of her brain's activity, and not her brain activity or consciousness itself. If the connection between brain and body becomes severed as a result of brain damage, blocking a person's ability to
move, then it's no longer valid to infer anything about what his body language shows about his conscious thoughts. It certainly is invalid to fully assume that a person has no conscious life because he is completely paralyzed.
Of course, we rarely have to worry about this problem, but with patients in
vegetative or related states, it is a real concern, and one that no behavioral
rating scales, however standardized or carefully constructed, will ever solve.

Adrian Owen, my primary supervisor during my PhD work and then for
many years my boss as I became a research scientist, has spent much of his
career searching for better, more direct ways to gauge the level of inner life
in such patients, with my occasional assistance. A charismatic ginger-haired
man, full of humor and uncannily resembling Vincent Van Gogh, Owen has
used brain-scanning as a powerful tool to probe the level of consciousness
in vegetative patients. These emerging techniques have proven, for some patients, to be far more sensitive than any based on behavioral observation.

One recent large-scale study he organized involved 41 patients who were
all in either a vegetative or a minimally conscious state. Owen and his team
placed patients in the fMRI scanner and presented them with various kinds
of sounds via earphones. The sounds were distinguished by their complexity.
First there was random noise for the patients' brains to process. Then came
normal speech, via standard sentences—for instance, "Her secrets were written in her diary." Finally, and most tricky of all to process, were sentences
containing ambiguous words—for instance, "The *shell* was *fired* towards the
tank."

By seeing which brain regions activated for each condition, and comparing this with the way that the brains of normal, healthy subjects light up on
the same tasks, he would thus have four possible levels of cognitive processing distinguishable by brain-activity patterns: inability to process any sounds,
ability to process random sounds, ability to process linguistic sounds, and
ability to process the meaning of words. Surprisingly, two patients who had
been firmly classified on behavioral scores as being vegetative were able to

pass the highest level and process the meaning of the words they heard, as shown by their brain activity. So they were probably minimally conscious instead—or better. More exciting still was Owen's discovery that the extent to which these patients passed the sound tests was a strong predictor of how much they would recover six months down the line.

This research is groundbreaking and may in future years become an important tool in the clinician's standard arsenal of assessing the level of consciousness in such patients. It should also give hope in certain situations, backed by good evidence, that a future improvement in symptoms is likely. Recall from Chapter 3 that understanding the meaning of words—a skill unavailable to those under general anesthesia—is one of the simplest functions requiring consciousness. Therefore, showing brain activity for meaning in this experiment is a solid clue that consciousness is present. It is also a sensible reason for why patients who pass this part of the test go on to recover further degrees of awareness.

But such data do not definitively show that these patients are conscious because it is too indirect and circumstantial. It is technically possible that consciousness is not present in these patients, even if meaning is appropriately processed. To answer such a vital question definitively, more convincing evidence would be needed.

This is exactly what Owen and colleagues provided in a landmark study using a somewhat different approach. Owen reasoned that if these patients could demonstrate volition, if they could choose to follow a complex command that he gave them, then that would provide unbreakable evidence that they were indeed conscious, regardless of what the clinicians' diagnosis was. He was able to test this idea in 2006 with a twenty-three-year-old woman who entered a vegetative state following a road traffic accident. She had already passed the previous sounds test with flying colors, showing robust activation to ambiguous words. But then Owen asked her while she was again in the scanner to carry out one of two tasks. Sometimes he would ask her to imagine playing tennis, and at other times he asked her to imagine navigating around all the rooms of her house (recall Chapter 1, when my friend Martin Monti carried out a similar experiment on me). When Owen looked at her brain activity for the two commands, her brain responses were indistinguishable from those of normal, healthy controls performing the same imaginary tasks. The only possible conclusion he could draw from this finding was that the doctors had misdiagnosed her, that she wasn't in a vegetative state, after all, but was in fact clearly conscious. Following this scan, she continued steadily to recover, and within a year she was able to answer yes-or-no questions by pressing a button with her foot.

Such a discrepancy between the standard clinical diagnosis and the brain-scanning results is dramatic, if relatively rare. In a later study, Owen, Martin Monti, and colleagues showed that only about 17 percent of apparent PVS patients could successfully show appropriate brain activity in response to the imagination commands. Still, this experimental approach shows that brain scanning can at times be a far more sensitive way of gauging levels of consciousness in these severely brain damaged patients than any bedside assessment by the doctor. Owen also believes that this result shows that some seeming PVS patients may already have leapfrogged over the intermediate minimally conscious state and have an almost normal level of consciousness—but they are simply locked inside their heads with no way of demonstrating awareness.

Because an fMRI scanner is a very expensive, nonportable machine, Owen's group has recently been testing the use of EEG instead. EEG is a more practical clinical tool, since it is far cheaper than MRI and can be taken to the bedside—an important factor for such patients, who are difficult to move. Although picking up a reliable signal with this technique poses a challenge, Owen and colleagues have nevertheless shown by a similar command-following protocol that it can be used to demonstrate that some apparently vegetative patients are indeed conscious.

COMMUNICATION BY BRAIN ACTIVITY

Such methods open up the possibility that, even though some of these patients give all outward appearances of being robbed of consciousness, not only are they clearly aware, but we might even be able to communicate with them via the brain scanner. When Martin Monti piloted his fMRI experiment on me, I answered yes-or-no questions by imagining moving around my house or playing tennis. He would then read off my answer in the brain activity the scanner picked up. Exactly this technique was later used on one patient, a young Belgian man, who had, it was assumed from outward appearances, been in a vegetative state for five years following a traffic accident.

When he was asked about the name of his father and other personal questions, he answered robustly and correctly via his selective brain activity to the first five questions. This was a striking result, completely at odds with his PVS diagnosis.

It's important to emphasize in all these studies that, as usual, a negative result on scans is hard to interpret—for instance, if a patient fails to generate the appropriate brain activity, she might just have moved too much in the scanner, which in itself spoils results. However, passing this test clearly reflects consciousness, and as methods are refined, it should become easier both for patients to prove they are aware and for them to communicate.

Although none of these methods are close to being a conventional clinical tool, there is every reason to assume that the situation will change in the next five years or so. Having a standard, brain-based method by which patients could express their thoughts and wishes, despite being completely unable to move a muscle, would give them a much-needed voice and provide hope and reassurance to their anxious families.

CHECKING THE INTEGRITY OF CONSCIOUS NEURAL HIGHWAYS

The methods described above essentially seek a lucky glimpse of conscious communication in the peaks and valleys of flittering brain activity. But what if the patient is having a bad day when scanned—for instance, spending much of this time asleep, when in fact he is usually very conscious? Or what if the patient really is only partially conscious, but all the neural architecture is present for a solid recovery? There is certainly mileage in applying less direct neural investigations that bypass questions of online consciousness or communication. Indirect methods with the most potential for these patients include those that examine whether the brain is sufficiently physically wired to support consciousness, or whether regions are still communicating with each other, in a way that reflects awareness. These methods, although less sexy, could nevertheless prove more robust than other techniques that rely on the patient obeying instructions, and they might also offer more reliable predictions of recovery.

And, given the level of maturity of the science of consciousness today, it is certainly possible to exploit existing knowledge about consciousness and the brain in order to make accurate diagnostic tests of PVS.

First, it is well known that the thalamus, that central relay station in the brain, has a critical role in consciousness. Although it might not be as important as the prefrontal parietal network in supporting the richness of our various experiences, its normal functioning, as a means of transmitting and combining information freely throughout the cortex, is necessary for awareness to occur. Significant damage to the thalamus, as seen in anatomical scans, is an obvious indicator that consciousness will be lacking and recovery unlikely. But even if the thalamus seems intact, that doesn't mean that its connections with the rest of the brain are still functional. So Davinia Fernandez-Espejo, Adrian Owen, and colleagues recently used a relatively novel MRI scanning technique, diffusion tensor imaging, in order to examine how well the thalamus was still connected with other regions in such patients. It turned out that the integrity of these fiber pathways was an excellent diagnostic measure of whether patients were classed as PVS or minimally conscious, securing a 95 percent accuracy. This is a considerably higher success rate than is typically reached by means of standard clinical behavioral assessments, which, you may remember, have a lamentable 43 percent misdiagnosis rate.

Looking directly at the pathways of the prefrontal parietal network is another clear research goal, emerging out of the large bank of evidence arguing that this is the most central set of brain regions for consciousness. Using a complex neuroimaging technique that assesses how one brain region causes activity in another, Melanie Boly and colleagues were able to show that, while in PVS patients the prefrontal cortex was no longer influencing regions toward the back of the brain, in minimally conscious patients and normal subjects there was a more robust, active link between the regions, with information flowing both away from and toward the prefrontal cortex.

Another principled approach, this time springing from information integration theory, uses the TMS-EEG technique mentioned in Chapter 6, where an initial burst of cortical stimulation, initiated by the TMS machine, is then monitored by EEG for its spread and duration. A prolonged bouncing of activity throughout the cortex is found in awake, aware individuals, while only a local, short-lived response is present when we're in a dreamless sleep. Although this research is in the early stages, the approach is also being used to measure the level of residual awareness in vegetative patients.

THE DIFFICULTY OF REPAIRING HUMPTY DUMPTY

Accurately finding ways of judging whether a vegetative patient is conscious is always only half of what's on the mind of the patient's loved ones. The family members also want to know if the patient will recover, and whether there are any treatments that will aid the healing process. Here, the news is not so encouraging, although there are a range of potential treatments.

For instance, Nicholas Schiff and colleagues carried out a procedure known as deep brain stimulation on one thirty-eight-year-old patient who had previously been in a minimally conscious state for six years. This technique involves an operation to insert electrodes deep into the brain, and in this case using those wires to continuously stimulate the thalamus for 12 hours each day. In this way, not only was the thalamus reactivated, but so were its many connected regions, including the prefrontal cortex.

Before the surgery, there were only the barest signs of awareness, with the patient occasionally able to follow verbal instructions, but unable to utter a single word himself. Almost immediately after the electrodes were turned on, the patient seemed to wake up far more effectively; he could now keep his eyes open and turn his head toward a spoken voice. This was dramatic in itself, but the patient continued steadily to improve. By the end of the study six months later, he was not only uttering single words for the first time and naming objects, but sometimes even saying short sentences. Whereas before he'd had little motor control and wasn't even able to chew food, now he could hold a cup to his lips and feed himself three meals a day. These are all remarkable improvements, but it's important to emphasize that, despite such dramatic progress, the patient was by the end still very ill, a shadow of his former self. It's also unclear whether this procedure would work at all on PVS patients, who, unlike this man before his operation, show no obvious signs of consciousness.

A different and controversial approach, this time deliberately directed instead to PVS patients, simply involves dosing these patients with medication. Ralf Claus and colleagues have provided provisional evidence that, for some vegetative patients, a common sleeping drug, zolpidem, can paradoxically allow them to wake up for about 4 hours at a time. It's almost as if their ability to sleep and wake has gone totally askew and this drug may temporarily reset and recalibrate the system. From showing no signs of consciousness beforehand, one PVS patient a few hours after zolpidem was administered could verbally answer questions appropriately and even perform simple calculations. Although other research groups have also shown that zolpidem can indeed improve conscious levels in such patients, it seems that only a small minority, about one in fifteen participants, will have any kind of benefit from the drug.

Occasionally, patients with few signs of consciousness can spontaneously recover on their own, even after years. In one famous case, a man named Terry Wallis suffered from massive brain damage and a coma after being thrown from his pickup truck. A few weeks after the accident, he was judged to be in a minimally conscious state. He remained in this state for nineteen years, until one day, out of the blue, he appeared almost normally conscious. Three days after this, he was already able to speak almost normally. This sounds, on the surface, miraculous, but when you start examining the details, you find that the situation isn't nearly as encouraging as it at first appeared. Wallis had severe amnesia, poor working memory, an inability to plan or strategize, impulsive behavior, and a changed character, now far less mature than his nineteen-year-old self at the time of the accident. In line with this, his brain scans revealed severe brain shrinkage, even though there was some promising evidence that new pathways were growing, which might have explained why he woke up in the first place.

With all these patients, a good prognosis critically depends on how much brain damage has already occurred and what caused the patient to be in this state in the first place. Returning to the case of Terri Schiavo, despite the hopes and prayers of her parents, her brain shrinkage was so severe that there was, sadly, very little brain left (see Figure 9), and certainly no hope that any pill or stimulator would fix her.

For the vast majority of these PVS patients, or indeed anyone who suffers from some form of serious brain damage, probably the best future hope for effective treatment is via some form of stem cell therapy. Theoretically, stem cells injected into the brain could turn into new neurons to reverse the extensive brain damage. At present, though, there are both real dangers to the treatment, with stem cells tending to form tumors, and great difficulties in rebuilding large sections of cortex. And even if, in the decades to come, stem cell therapy did mature and become a viable treatment, it might well be that patients would emerge with most of their old memories, skills, and personalities wiped out from the initial brain damage. They would be disturbingly reborn, with a new consciousness and character to grow, piece by piece. It's hard to imagine what extra challenges such a patient would face, when even in normal development and in our daily lives we so frequently struggle to find a mentally healthy path through our experiences.

8

Consciousness Squeezed, Stretched, and Shrunk

Mental Illness as Abnormal Awareness

SHARP FRACTURES IN AWARENESS

My wife and I have been together for thirteen years. In some ways, because of her fierce emotional warmth, her busy intellect, and the fact that we think alike on many topics, it has been very easy to be with her. In other ways, through absolutely no fault of hers, it has been, on occasion, a struggle. My wife suffers from one of the various forms of bipolar disorder and, like many people plagued by this illness, she endured many exasperating years as the psychiatrists toyed with a sequence of alternative diagnoses before landing on the current label. They then prescribed an even longer sequence of largely ineffective medications, and we reacted to each new drug with the increasingly tired cycle of hope, frustration, and despair.

When my wife is ill, she is usually "down": She will sleep much of the day, profess to be perpetually lethargic, and be globally, profoundly lacking in motivation—sometimes to such an extent that she can hardly move her body. She can't make decisions, can't recall normally vivid memories, and will fail to understand topics that in normal circumstances she'd grasp with ease. She will feel either numb or terribly distraught. Her thoughts and feelings will bend heart-wrenchingly away from rationality. They will converge on the view that she is the most ugly person alive (when in fact she is very good looking), that she is incredibly fat (she has a BMI of about 21), that she is utterly stupid (she holds a PhD in genetics, along with a clutch of other degrees, from both Oxford and Cambridge Universities), and that she is the most unkind, unpopular person in the world (she in fact has a large set of friends, and is very giving toward them).

Far less frequently, she will be "up": She will sleep little—perhaps skipping bed for a night or two. She will feel high, disinhibited, excited, and easily excitable, almost as if she were drunk. She will have lots of energy and start various new projects, some of which, under normal circumstances, she would think were a complete waste of time. She will make lots of connections between ideas, rather like forming new, insightful chunks, but at their extreme form, the connections will make no sense, and the ideas may be somewhat absurd and, in hindsight, embarrassing.

Of course, illnesses like bipolar disorder are tremendously complex. But I've been struck at times by the extent that her symptoms could be explained by a warping of her consciousness. When she's within a depressive episode, it's as if there isn't enough consciousness to go round; she's perpetually tired, and she lacks the awareness to realize the extreme irrationality of her beliefs about herself. When she's in a manic episode, it almost seems as if there's too much consciousness for her brain to cope with; she never gets sleepy, her melting pot of ideas is boiling too hot and bubbling over the rim, and her "innovation machine" of a conscious mind repeatedly spews out spurious insights.

She was prescribed many of the standard drugs for depression and bipolar disorder—for instance, selective serotonin reuptake inhibitors (SSRIs), to raise her mood (by raising her serotonin levels), and mood stabilizers, like lithium (whose effect no one really understands), to keep her on an even keel. Although some of these drugs made her physically very ill, and one or two made her mental health plummet frighteningly, none helped her bipolar symptoms (a somewhat common experience in patients like her). Eventually, we were lucky enough to chance upon a particular stimulant medication. This drug, within a day, removed the bulk of her depressive symptoms. It's not perfect, and may not be suitable for many other bipolar patients, but it is orders of magnitude better than the more conventional medications she tried and allows her to lead a far more functional life than before.

But why should a drug that promotes wakefulness have such a profound effect on one's feelings of worthlessness? And what other mental illnesses might benefit from being perceived as more closely connected with consciousness than previously thought?

The World Health Organization (WHO) estimates that up to a quarter of all people around the globe are affected by mental illness at some point in their lives—with anxiety and depression the most common conditions. Suicide now ranks as one of the leading causes of death in young adults. Unlike other major illnesses, such as cancer and heart disease, which tend to occur later in life, mental illness is most likely to emerge in adolescence or early adulthood. This means both that the pain to the patient can be more long lasting and debilitating and that its economic burden to the state is larger.

Although it's difficult to estimate the loss to the economy due to mental illness, the World Economic Forum recently made an attempt at this. Taking all the relevant factors into account, such as direct costs to treat the illness, work-hours lost to disability, and so on, it calculated that in 2010 the global cost of mental illness was around $2.5 trillion—a staggering amount, but likely to increase dramatically over the next twenty years. Mental illness will thus soon account for the majority of the world's economic burden for all noncommunicable diseases (which also include conditions such as cardiovascular disease, chronic respiratory disease, cancer, and diabetes). The entire global health budget is only about double the current mental illness global economic burden, though a tiny proportion of this is spent on mental health. In fact, mental illness must surely be the most underfunded sector of health care by far, when the true costs of mental illness are considered. Therefore, one of the most sensible policy changes that could be undertaken to positively affect economic growth figures would be to focus on improving mental health in the populace.

Perhaps part of the reason for the political neglect of mental illness is the legacy view that these aren't real illnesses, as well as the assumption that psychiatric conditions are too complicated to treat effectively. While the first assumption is inaccurate and damaging, there is some truth to the second position. As my wife and I know all too well, the decades of research have hardly scratched the surface in our understanding of mental illness. This is probably because many of these conditions are caused by a large set of genes that interact with environmental events in extremely complex ways. But that doesn't preclude new perspectives from shedding light on these profound sources of suffering. And even minor progress in treating mental illness, a disturbingly common condition, could lead to a dramatic improvement for society, both from its soothing reduction of individual torments and from the potential to make the economy more productive and efficient.

In this chapter I consider the idea that almost all mental illnesses can be rewritten as disorders of consciousness. This is a strategy that may engender useful, novel perspectives on these conditions, as well as exciting new routes for treatment. I'll also suggest in the epilogue to follow that insights from our emerging knowledge of the nature and purpose of consciousness can help explain and alleviate the emotional struggles we all face on a day-to-day basis.

AUTISM AND OVER-CONSCIOUSNESS

From the perspective of consciousness and psychiatry, autism stands out as a special, fascinating syndrome. Although historically seen as a disorder defined by social impairments, it is rapidly being rewritten as a condition caused by an overabundance of awareness.

A developmental disorder that is apparent from early childhood, autism symptoms include a lack of understanding of the thoughts and emotions of others, poor language skills, and obsessive, repetitive behaviors. The majority of autistic children are classed as mentally retarded and have the condition for life.

As with most psychiatric conditions, autism is probably best viewed more as a cluster of symptoms than a single syndrome, and there is much variability in level and type of disability among autistics. For this reason, most people talk of autism spectrum disorders, rather than simply "autism." And this continuum includes people with Asperger's syndrome, who have some symptoms of autism, such as poor social skills, but who generally function well in life, and may have a high IQ.

The traditional view is that all autistic symptoms stem from an inborn, specific deficit in grasping anything remotely social—this is why autistics can't understand other people, and in pronounced cases don't learn the social activity of communication. Most theories along these lines, though, struggle to explain the full range of symptoms common in autism, such as repetitive behaviors, which seem to have nothing to do with a lack of social skills. And this traditional view seems out of sync with emerging evidence that many severely autistic children can show marked improvements in social and verbal skills if they participate in an intensive behavioral intervention program centering around gentle social encouragement through play, such as the Early Start Denver Model or the Son Rise Program. This raises the strong possibility that these social and communication problems, far from being the central cause of autism, are a reversible side effect of a deeper difference between autistics and others.

One emerging suite of theories on autism suggests that this disorder is centrally defined not at all by a poverty of mental skills, but instead by the excessive richness of information these people experience. In other words, in some sense autistics have an overabundance of awareness, and all their symptoms are merely their way of dealing with this supercharged consciousness.

But how does this fit with the view that autistics are so poor at processing information that they are regularly classed as mentally retarded? The simple answer is that it doesn't fit, because most autistics aren't really mentally retarded at all. For a start, the Asperger's subbranch of the autistic spectrum involves individuals with a normal, or, more usually, a raised IQ, sometimes markedly so, such that many physicists, mathematicians, and engineers probably have some autistic traits. Simon Baron-Cohen, a world expert on autism, and Iain James have speculated that certain prominent figures, including Albert Einstein and Isaac Newton, suffered from Asperger's syndrome. Einstein, for instance, was very developmentally challenged in learning to talk, was a loner as a child, and would obsessively repeat sentences until about the age of seven.

Some autistics really are severely mentally disabled—for instance, they might be unable to go to the toilet by themselves—but I don't believe this is *necessarily* because they lack the mental resources to perform such a task. Trying to assess a child who has poor language skills and an aversion to novel activities is remarkably challenging, but increasingly scientists are critical of this mentally retarded label for autistics, and this may well be another case of researchers jumping to the conclusion that poor performance on a test must mean inability to carry out a related everyday function.

There are two popular, but very different, ways of measuring IQ: the Wechsler Intelligence Scales, which use a shotgun approach, giving subjects a large range of different tests, including various language tasks; and the Raven's Progressive Matrices, a single nonlinguistic test that involves finding the logically correct option to complete the hole in a patterned grid. Unsurprisingly, many autistic children are rated as severely mentally retarded on the more conventional Wechsler IQ scale, with its heavy language component. But if you give autistics the Raven's matrices test instead, suddenly their IQ jumps. Their scores are, on average, 30–70 points higher than they were on the Wechsler test, and as a group they appear at least as intelligent as normal children.

Thus, autistics generally aren't mentally retarded, and their social and communication problems can be dramatically improved with the right behavioral therapy. But what evidence is there that they are actually more conscious than normal people? Actually, there is a wealth of data from a wide range of sources.

A small subset of autistic people will exhibit isolated islands of incredible ability. For instance, there's Stephen Wiltshire, who can draw a landscape with stunning accuracy after just a single viewing. Or there's Derek Paravicini, who may not know how to hold up three fingers, but despite his blindness is a highly accomplished pianist who has filled concert halls and can play a tune perfectly after hearing it just once. And there's also Daniel Tammet, already mentioned in this book (Chapter 5), an autistic man who can perform fiendish calculations and memorize gargantuan streams of numbers.

Until recently, it was thought that only 10 percent or so of autistic individuals had some form of superior skill. But increasingly, it's becoming clear that most have superior abilities in a range of perceptual and analytical areas.

From a biological perspective, in many ways autism is the opposite of schizophrenia. The two conditions share many of the same gene abnormalities, but while schizophrenics will have one variant of a given gene—say, involving a deletion of a section of DNA—autistics will have the opposite variant—for instance, with a duplication of the same section. And while schizophrenics show a slowed brain growth in childhood, autistics have an accelerated brain growth compared to normal children.

Many autistics seem to perceive the world with more detail than the rest of us and can exercise a highly focused and sustained attention. But being flooded by so much detail can be stressful and overwhelming. Autistics tend to hate noisy or busy scenes; while sharp, unexpected sounds cause most people's consciousness to be taken over by this new event, in autistics the intrusion is particularly pronounced. Some even report a low-grade form of synesthesia, where a particularly loud noise will induce a brief visual flash,* as if the intensity of the sound was so powerful in them that it spilled into other senses. This inadvertent sensory mixing also hints at the autistic's marked propensity to combine information.

If autistics have a wider consciousness than others, from the main thesis of this book it would follow that they would then also have a more patterned, structured mind. This, I would argue, is one of the hallmark features of autism, and one of the most common symptoms, from the most disabled autistic child to the most towering genius classed as having Asperger's syndrome.

Many autistics, in order to compress their overflow of conscious detail and reduce the related stress, find comfort in crafting structure inside their minds and in their surroundings. This is why they may build blocks in carefully crafted stacks, or develop many rules and rituals. These activities are a desperate attempt to reduce the novelty and chaos in their lives and replace them with reassuringly regular, compressed patterns. It is also why they tend to seek highly ordered hobbies, such as mathematics, being a human calendar, and so on. There is in fact only a small range of superior skills that autistic people adopt, and these are almost exclusively related to logically crisp, hierarchical patterns of meaning. Many of us may nurse a deep hunger for patterns and structure, as exemplified by the sudoku and crossword puzzles we enjoy. But for autistic people, this need is tangible and chronic. For them, attention to such structures, and the habitual application of chunking, is a lifeboat they latch onto so as not to drown in a sea of conscious information. In many cases, it may also be the main source of pleasure and security in their lives. For instance, in his autobiography *Born on a Blue Day*, the autistic Daniel Tammet eloquently wrote, "The pages of my books all had numbers on them and I felt happy encircled by them, as though wrapped in a numerical comfort blanket."

---

*I suspect that synesthesia is more prevalent in autistics than in the general population, although I don't know of any studies that have as yet looked into this.

I suspect that for some autistics, delayed language development or impaired social skills partly arise because these people are initially drawn to other, more patterned aspects of the world, and they have little inclination to learn certain linguistic or social skills of which they are probably very capable. Tammet, for instance, despite his autism, has since his mid teenage years made a concerted effort to learn the social rules. He seemed a particularly amiable, helpful person whenever I met him. He can also speak about ten different languages, and revels in the logical structure hidden behind words. It might also be, though, that many autistics feel easily saturated and upset by the social world, which can be so much more chaotic, random, unpredictable, and uncontrollable than nonsocial, more logical domains. In other words, autistics may initially shy away from social situations because of the stress of an overflow of a form of information that they can't effectively compress.

Although the wider mental landscape that some autistic individuals possess makes the term "disabled" a questionable label here, if the condition is combined with undue suffering, then there are ways to help. Behavioral therapies are already exploiting the idea that an inability to understand the emotional and social worlds is neither a central nor an immutable feature of autism. And although there has been limited success so far in pharmacologically treating autism, one new drug, arbaclofen, has produced provisional yet promising results. And the way that this drug is thought to work may be in line with the suggestion that autistics have too much consciousness. Normally, our brains are awash with some neurotransmitters that encourage neurons to fire and others that turn down the signal, dampening down neuronal activity. It is thought that autistic people have an imbalance of these chemicals, having too much of the main neurotransmitter that ramps up activity (glutamate) and not enough of its nemesis, gamma-amino butyric acid (GABA), which suppresses neuronal firing. Arbaclofen acts to restore this balance, providing autistics with a more normal, less overwhelming sense of consciousness, and as a result social skills develop more rapidly.

UNHEALTHY SLEEP, UNHEALTHY CONSCIOUSNESS

While autism may be unique in being a disorder of an overabundance of awareness, most other psychiatric conditions may be caused by a shrunken and skewed consciousness. And for a surprising proportion of these patients, part of the cause of a diminished awareness is a simple source that could be easily corrected: sleep disruption.

I feel very lucky that I've never had a mental illness. Whether or not it's related in my case, I'm also, as long as I keep the espresso drinking down, a solid sleeper. But it doesn't take much sleep deprivation for me to catch a glimpse of what some mental illness sufferers endure. In my student days, I had my fair share of nights with little sleep because of looming assignment deadlines, or more social reasons. If, on rare occasions, there were two nights in a row with less than a few hours' sleep, I felt like a different person—very much for the worse. I saw myself as temporarily half-senile. I seemed to be suffering from amnesia for the little, yet important, details, like computer passwords or the names of my friends. I struggled to perform simple actions, such as cooking a meal, and my mind generally seemed to be in a perpetual fog. I became far more anxious, more prone to stress, and less confident than usual. Most striking of all, two days with little sleep and I approached the edges of hallucination. I didn't hear alien voices in my head, see imaginary people in front of me, or anything else that dramatic. On a frighteningly regular basis, though, I could no longer be sure whether or not my name had been mentioned in a nearby conversation between complete strangers, or whether the pattern in the carpet had started moving of its own accord. Unlike those with a real mental illness, I was constantly reassured by the fact that one good night's sleep would instantly cure me. But for so many people, that is simply not an option.

Sleep and dreams are a pervasive and bizarre part of our lives: We spend about a third of each and every day semi-comatose in our beds, deliberately reducing consciousness either to a near void of experience or to surreal, disconnected vignettes locked inside our heads. The world's most popular drug, caffeine, is enjoyed mainly for its ability to ward off feelings of sleepiness. Deprive us of the land of nod for just one night and we immediately develop profound impairments in cognition. Deprive us for two or more nights, and we may begin to show transitory signs of psychosis. After 48 hours, we will find it almost impossible to fight the desire to sleep. This is in surprising contrast to something as seemingly vital as food.

Brought up in a religious Jewish household, I am no stranger to fasting. The Day of Atonement, Yom Kippur, involves a total fast of food and water for 25 hours or so (though no such sleep fast is included). I remember one Yom

Kippur when I was eleven. Under the rules, I needn't have fasted until I had passed my thirteenth birthday. At this stage, my religious views were already starting to dissolve into atheism, but ever the scientist, my curiosity for mini-personal experiments was already quite apparent, so I insisted on fasting to see what it would feel like and how I would cope. By midafternoon, I'd already been fasting for about 19 hours, with no water or food at all. At this point, with a group of friends, I secretly ducked out of prayers to play a game of soccer in a deserted hall on the far side of the building. Running around for an hour is not the most sensible option when you are probably somewhat close to dehydration, but I seemed to be no worse for the activity, and was struck when the fast was broken that evening during dinner by how easy it had been to go without sustenance for an entire day.

In striking contrast, I don't think I've ever gone without sleep for a full 24 hours and not had to struggle desperately by the end of it to stay awake.

Experiments in rats indicate that total sleep deprivation for up to a month is consistently fatal. Almost all animals sleep, including insects. Most scientists believe that our ability to nod off, and especially to dream, is a key ingredient in effective learning and memory. Single neurons, it appears, can become overused and "tired" in the day; they need a period of reduced activity to reset themselves and become ready for a new period of learning tomorrow. Sleep is one of the prices we pay for flexible, information-hungry brains.

Therefore, if we want a trim, smoothly running consciousness, we need healthy periods of peaceful slumber. This connection between sleep and health might not be the most earth-shattering scientific revelation, but nevertheless it is a fact we all too readily ignore. In the teeth of our stressful, busy modern lives, we all regularly underestimate how delicate sleep can be and how profoundly it can affect our mental health.

Humans probably top the animal kingdom in the sheer range and prevalence of sleep problems we collectively endure—just as we top the animal kingdom in the range of psychiatric and neurological illnesses we are prone to. Common sleep disorders, unfortunately, are another of those payments we have to make for such a complex brain and such a broad, supercharged consciousness. It's been known for decades that sleep problems are associated with a host of psychiatric conditions. But the conventional wisdom was always that sleep problems were a symptom, rather than a cause, of psychiatric disorders. Now that assumption is increasingly being turned on its head.

On the surface, determining whether sleep difficulties cause, or are the effects of, mental illness is a tricky issue to disentangle: A person on the cusp of a depressive episode might feel a spike in stress because of her debilitating feelings, and the stress might keep her up at night, which will make her more stressed, anxious, and upset, and the downward spiral continues. But in fact, there are methods to bypass this confusion.

Sleep-related breathing disorders cause profound sleep disruption, usually without the slumbering person realizing there are any problems, even for many years. The most prevalent kind, sleep apnea, normally manifests as pauses in breath as frequently as every few minutes throughout the night as the windpipe becomes blocked. This pseudo-suffocation forces the person to wake up for a moment, but normally not long enough to remember the event the next morning. Crucially here, the cause rarely has to do with the brain, and instead usually arises from mechanical deficiencies in the throat. Paul Peppard and colleagues examined whether sufferers of this condition would be more likely to have depression, and this is indeed what he found, with the risk of depression related to the severity of the breathing problems. In other words, merely having chronic, poor-quality sleep can induce depression.

Closely related to this is seasonal affective disorder (SAD), a form of depression that rears its debilitating head in the winter months when there is less daylight around. Its sufferers feel generally sleepy, lacking in energy, and low in mood—almost as if they have half entered into hibernation (which might well be what is occurring, as an evolutionary remnant of some earlier mammalian form). The main, commonsensical treatment for this condition is light therapy, where patients sit near specially designed lamps as a substitute for the sunlight they are missing. The effect of this treatment is to raise alertness levels, which at the same time helps blast away those blue feelings.

Attention deficit hyperactivity disorder (ADHD) is a childhood condition with even more obvious links to sleep and consciousness. Admittedly, this condition is, in some people's eyes, a rather amorphous and vague catalog of symptoms, which allows for dangerous overdiagnosis, medicalization, and management via potentially addictive drugs. Still, it's clear that a subset of children labeled as ADHD have a serious disorder, with a genetic basis, that makes them disruptively impulsive, hyperactive, and unable to concen-trate, with a shriveled working memory. The fact that working memory, the mental playground of consciousness, is diminished is itself a good indicator that consciousness is lower. And sleep is likely to be one of the causes of this, as again shown by sleep apnea.

David Gozal and colleagues found that five- to seven-year-old children were considerably more likely to have symptoms of ADHD if they had sleep apnea. Although the most pronounced ADHD children didn't show this re-lationship, possibly because their symptoms were already being generated by underlying genetic abnormalities, their sleep was still far from normal: Their rapid eye movement (REM), or dream-period, sleep was significantly curtailed and commenced later in the night than in normal children. This portion of our night's sleep is the most important for learning and memory. If we have a poor night's sleep, the proportion of our REM sleep the next night increases to catch up. Disturbances in REM sleep have also been spot-ted in depressives and schizophrenics.

If you've ever witnessed a child at the end of the day who is overtired but also strangely excitable—and more likely to misbehave—this emerging pic-ture of ADHD as akin to a chronic lack of sleep may make considerable sense. ADHD sufferers can report almost feeling like zombies, as if they are only semiconscious but are automatically doing all they can to overcompen-sate for this awkward state. Supporting this view, the prefrontal parietal net-work, the key network for consciousness, is chronically underactivated in ADHD children.

You might think that to combat hyperactivity, doctors would prescribe valium or low-dose sedatives to cajole these patients' minds into a quieter, more serene state. Instead, as if in tacit acknowledgment that these patients are permanently overtired, stimulants such as Ritalin are ubiquitously used for ADHD.

As some of my wife's symptoms suggested, bipolar disorder can compete with ADHD for its ties with sleep. For instance, manic episodes are far more likely if a bipolar sufferer is forced to miss a night of sleep, or flies into a dif-ferent time zone, thus unbalancing the sleep-wake cycle.

In fact, in almost any major mental illness, the ties between disturbed sleep and psychiatric symptoms are becoming increasingly clear. For instance, abnormalities in genetic and molecular processes that regulate sleep have been linked with unipolar depression, seasonal affective disorder, bipolar disorder, mania, panic disorder, post-traumatic stress disorder, obsessive compulsive disorder, ADHD, and schizophrenia.

If many of these conditions relate to chronic under-awareness in normal waking life, then it stands to reason that stimulants, whose purpose is to increase arousal, may be of benefit in relieving symptoms. As mentioned above, this is the conventional treatment for ADHD in the form of Ritalin, but what of other conditions? One of the clues that led my wife to stimulant medication to treat her bipolar disorder was that she had clearly been, for many years, self-medicating on lots and lots of coffee during the day. This was the only method she had to feel better. It turns out, perhaps unsurprisingly, that coffee has a rather robust protective effect against depression for many women (so far, the main studies have not included men). For instance, Ichiro Kawachi and colleagues demonstrated, quite convincingly, that coffee intake prevents suicide. And another recent study by Michel Lucas and colleagues showed that increased coffee drinking is associated with a lower incidence of depression. Psychiatrists seem slowly to be turning toward the medical forms of stimulants as effective alternative treatments for many psychiatric conditions, but an accelerated push in this direction may be warranted in order to yield more breakthroughs.

Another emerging weapon in the clinical arsenal to correct sleep problems and raise alertness in mental illness is light therapy. While it's well known that this treatment is very effective against seasonal affective disorder, it has also been shown to help in all manner of psychiatric conditions, including any form of depression, ADHD, and even Parkinson's disease and other forms of dementia.

In short, a quiet revolution is building in psychiatry, where sleep abnormalities are viewed as an important potential cause of mental illness rather than another symptom. And treatments are starting to target the fact that in these patients sleep quality needs to be improved and daily arousal levels increased.

WORKING MEMORY NOT WORKING

While the evidence is mounting that a paucity of wakeful consciousness can cause an array of mental illnesses, the question remains as to why exactly this could turn some people into psychiatric patients, as opposed to simply feeling tired. As the model defended in this book asserts, a reduced awareness relates closely to a reduced working memory capacity. This in turn can lead to less mental control, a lower ability to innovate your way out of trouble, and a smaller chance that you'll notice when your thinking or behavior may be moving in dangerously wrong directions.

What is the evidence that sleep problems actually shrink your working memory? The evidence for children comes from observing their natural day-to-day sleep quality and seeing how this relates to working memory. A poorer quality sleep does indeed impair performance on a range of working memory tasks. In healthy adults, you can be manipulative rather than observational, forcing an entire night of sleep deprivation and then monitoring the effects both behaviorally and in the fMRI scanner. Michael Chee and Wei Chieh Choo carried out such a study and found that a lack of sleep created both working memory problems and a less efficient prefrontal parietal network than normal. In this study, healthy volunteers in their early twenties missed just one night's sleep and effectively had brain activity that resembled that of seventy-year-olds. Robert Thomas and colleagues, studying adults with sleep apnea, found very similar results to the sleep deprivation study just mentioned. Fragmentary sleep in this patient group was associated both with a slowing of performance and reduced accuracy in a verbal working memory task, as well as a reduction in prefrontal cortex activity.

There is also circumstantial, though no less intriguing, evidence to link sleep issues with a diminished working memory. Just as light therapy helps wake up psychiatric patients, improving their symptoms in the process, in normal participants it boosts performance on a range of attention and working memory tasks.

So, from a broad range of sources, it's clear that sleep problems do indeed reduce working memory capacity, and, in line with this, the core cortical regions of consciousness are underpowered when sleep problems are present.

In ADHD children working memory levels are especially low. How do they cope with this dwarfish holder for conscious contents? They tend to do all they can to avoid tasks that will strain their limited conscious resources—they shy away from difficult school problems and flit between many tasks, because they can't maintain a single goal in working memory for long. And they behave in impulsive, sometimes violent ways, since it can require considerable conscious effort to suppress primitive impulses and act in a controlled, reasonable manner.

But it doesn't take much for control in any of us to be impaired because of a deflated consciousness, as indexed by reduced working memory space. In one striking experiment, by Baba Shiv and Alexander Fedorikhin, undergraduate students were given either a seven-digit or a two-digit number to memorize. While holding this number in working memory, they were shown their reward for participating—a cart of snacks. They could either choose fruit salad—a healthy but not particularly indulgent option—or a piece of chocolate cake, which looked delicious and sweet, but also fattening. Students who were rehearsing the difficult seven-digit number were far more likely to choose the chocolate cake. It's as if their working memory, filled up by the demanding seven-digit number, didn't have sufficient space left to carry out normal, mature control over their feeding habits.* If such clear results can be obtained in healthy adults, simply by remembering a longer number, it's not hard to see how ADHD children, with such a diminished consciousness, will behave far more impulsively and appear out of control.

Another psychiatric group marked by a considerably reduced working memory capacity is schizophrenia. In fact, this is such a pronounced trait that some theorists have suggested that diminished working memory space is the prime psychological cause of much or all of the various nightmarish symptoms that schizophrenics endure. And, like ADHD patients, schizophrenics also have a dysfunctional prefrontal cortex.

A reduced working memory and shrunken consciousness have many knock-on effects. Effortful, taxing tasks requiring a deep understanding or a firm control over oneself must seem an impossibly high mountain to climb for ADHD patients, so they simply avoid such activity. Schizophrenics, with their distinct suite of genetic abnormalities, react differently to the problem, not even having the mental space inside an episode to notice their cognitive limitations. Instead of developing strategies to avoid the information overload, as ADHD patients do, I suspect that schizophrenics embrace this saturation

---

*There is a clear relationship between chronic poor sleep and obesity, although the mechanism for this is unclear. I suspect that one part of the effect is the daily playing out of the equivalent of this chocolate-cake study, where poor sleep shrinks working memory, and thus mental control—and a less healthy diet ensues.

of mental details and constantly strive to discover the hidden structures within their broken conscious stream. Autistics also search for patterns incessantly, usually with great success, and in a way that may help alleviate their suffering. Normal people do as well, of course, but we are prone to jumping to conclusions and to succumbing to superstitious beliefs as we make too much of too little data. Schizophrenics, via their misfiring awareness and paucity of working memory space, take these flaws of insight and magnify them a hundredfold. The result is usually an elaborate, heartbreaking edifice of delusions and hallucinations that schizophrenics have no hope of correcting with their deficient pattern-spotting consciousness.

CHEMICAL CASCADES UNBALANCING AWARENESS

If a reduced working memory is a core feature of both ADHD and schizophrenia, what treatments can target this, boosting working memory and consciousness while simultaneously alleviating symptoms?

Drug treatments for ADHD tend to rely on the assumption that the prefrontal cortex is underperforming. For instance, the stimulant drug Ritalin acts by increasing available amounts of the neurotransmitter dopamine. This neurotransmitter is vital for normal prefrontal parietal network function. If healthy, mentally well subjects are given Ritalin, and then receive brain scans, it is found not only that their working memory performance improves (especially if they started with a low working memory capacity), but that their prefrontal parietal network functions more efficiently. Ritalin may therefore help ADHD sufferers by flooding the consciousness network with a brain chemical it somewhat lacks, and in this way ramping awareness, in terms of working memory capacity, back up to near normal levels.

Unfortunately, though, most of these stimulant drugs do not work at all for schizophrenics, who already have an overabundance of prefrontal dopamine (in fact, giving the ADHD drug of choice, Ritalin, to schizophrenics is a reliable way to create psychotic symptoms). For optimal working memory, you actually need a medium amount of dopamine in the prefrontal cortex—too much or too little, and working memory shrinks.

For many years it has been assumed that correcting the prefrontal dopamine imbalance is the key to treatment for both ADHD and schizophrenia. Indeed, most antipsychotic drugs used for schizophrenia are designed to reduce dopamine levels. Although treatments targeting dopamine levels largely seem to fit for ADHD, emerging research is suggesting that for schizophrenia, scientists have been centering on the wrong chemical all along. Instead, the evidence now points the finger at glutamate, the most common neurotransmitter in the brain, whose purpose is mainly to cause neurons to fire and maintain consciousness.

While autistics have a supercharged glutamate system, their genetic opposite, schizophrenics, have severely deficient glutamate signaling, which has a knock-on effect on dopamine activity in the prefrontal cortex. So while ADHD patients have a reduced working memory primarily because of a lack of dopamine in the prefrontal cortex, schizophrenics' similar working memory problems have their source in much broader chemical imbalances that limit proper conscious function throughout the entire system. This probably means that schizophrenics also have a badly disrupted attentional system, so that all manner of unhelpful, unwanted feelings and ideas can gate-crash into consciousness. This feature, combined with a shrunken capacity to store such information within awareness, ever more readily creates a fractured, disorganized information overload.

One intriguing clue that has helped to shift the research perspective away from seeing schizophrenia as primarily a dopamine disorder comes from anesthesia. Ketamine is a general anesthetic that removes consciousness by suppressing a key part of the glutamate system, and with less of this neuron-activating chemical pathway, the brain's activity dramatically drops. But if given in lower doses to normal people, ketamine can, for the short while that the drug flows through their brains, effectively turn them into schizophrenics, as well as reduce their working memory capacity. For instance, otherwise mentally healthy volunteers will start to hear imaginary popping sounds, or become convinced that the study is being run by aliens. If given to schizophrenics in these low doses, ketamine can greatly exacerbate all their symptoms, and patients will say it is like being back in an acute episode. So here's a drug that if given in low doses dampens neuronal activity in a certain way, but not quite enough to make people unconscious. But consciousness is nevertheless still reduced, working memory drops, and the result is that normal people temporarily turn into schizophrenics.

Therefore schizophrenia, with such vivid, active, disturbing symptoms, may on the surface appear to have little to do with a reduced consciousness.

But dig deeper, and on any level you care to examine—with a diminished working memory, prefrontal dysfunction, and glutamate at semi-anesthetized levels—it becomes clear that a reduced consciousness is exactly how you should characterize this illness.

A new perspective on schizophrenia is particularly vital because the sad fact is that the current crop of drugs given to schizophrenics, still centering on the dopamine system, are poor at helping these patients. They seem to improve matters not so much by removing the various delusions and hallucinations, but largely by acting as a partial sedative, so that the tormenting symptoms are blunted. As might be expected from drugs that are only targeting a secondary neurotransmitter imbalance (an overabundance of dopamine), rather than the underfunctioning chemical ringleader (glutamate), only about 40 percent of schizophrenics gain any benefit from them. In contrast, far more robust figures can be obtained when examining significant adverse side effects: Approximately 67 percent of schizophrenic patients report these, with excessive sleepiness one of the most common problems. There is even anecdotal evidence that some schizophrenics function better in the long term if they've never been prescribed antipsychotics.

What is needed instead are drugs that focus on the root cause—abnormal glutamate activity. If a medication could ensure normal levels of this neurotransmitter, it might also indirectly generate a more functional dopamine system in the prefrontal cortex and in the process restore normal consciousness and improve working memory. Although as yet only at the research stage, new forms of medication along these lines are being developed. For instance, Sandeep Patil and colleagues, of the pharmaceutical company Eli Lilly, have created a drug that targets a specific part of the glutamate system thought to be dysfunctional in schizophrenics (this drug's mechanism, incidentally, is largely the opposite of that of arbaclofen, a drug currently in trials to treat autistic sufferers). In early trials, schizophrenics showed a positive response to the drug, and it is hoped that these and similarly acting medications will soon arrive in the clinic and allow schizophrenics to lead a more normal life than is currently possible.

As a more general point, many mental illnesses are relatively poorly treated with medication. Developing new drugs is only part of the problem.

Drugs already exist that can individually target many of the neurotransmitters in the brain. But psychiatrists have little means to assess what is wrong with their patients on the level of brain chemistry and do not really know how the patient might react to a given form of medication before it is provided.

When accompanying my wife to her psychiatric appointments, I found that the main focus of the doctors each time was to reconfirm her diagnosis and then prescribe the standard medications for that imprecise condition. They would start with the most popular drug and, as each successive pill failed to help her in any way, work their way down the list, trying a new one every six months or so. Then, when they were beginning to exhaust single treatments, they would start combining drugs. I would ask the doctors what they knew about how each cocktail of drugs interacted on the brain's chemistry. They would tell us that the combination of drugs had not been shown to be harmful, but they believed that no one understood what happened to the brain when you combined such drugs. All they knew each time was that such combinations had been helpful for some patients.

These experiences illustrated to me where the main pressure point of imprecise neuroscientific knowledge lies. We are making great strides in our scientific understanding of the human brain, and there are some relatively clear cases, such as schizophrenia and the neurotransmitter glutamate, where a previous approach was missing the point, and a new, more accurate viewpoint may provide striking benefits. But, on the whole, neural complexities are utterly daunting. The more we learn, the more complicated the microcircuitry of neural communication appears. There isn't just the tricky issue of how one neurotransmitter will interact with others. On a smaller scale, single genes can have many different effects in the brain and can be turned on or off by small sections of RNA; other short sections of DNA, not part of any gene, and previously classed just as "junk DNA," can nevertheless indirectly change neurotransmitter function as well.

So clinicians in this field have an almost impossible lot. Their job is to treat the debilitating mental symptoms of the patient, but the true problem is happening somewhere in the brain, probably relating to a dysfunctional neurotransmitter system—though there could also be anatomical abnormalities, with brain regions not built quite right in the first place. And even if the neurotransmitters are faulty, the cause of this could be one of a hundred biomechanical possibilities.

Thus, on the one hand, it's entirely understandable that psychiatrists remain primarily concerned with uncovering psychological problems and making pragmatic, admittedly crude attempts to alleviate those symptoms with one or a few of their arsenal of possible pharmaceutical weapons. On the other hand, absolutely any attempt in the future to edge closer to the locus of the problem by detecting brain abnormalities in these mentally ill patients, and to use more targeted solutions based on those clues, would almost certainly be of great benefit.

Already there are important successes emerging in this direction as scientists try to untangle the dizzying complexities of the brain in areas relevant to psychiatry. Some of these results could soon find their way into the clinic to aid diagnosis. For instance, an increasing number of genes coding for dopamine and prefrontal function are being associated with psychiatric conditions. Such results reinforce the view that consciousness should be the main context by which investigators should search for clues to mental illnesses. Other genes have been discovered whose variants determine whether or not you will respond to SSRIs, or whether such drugs will even pass the blood-brain barrier and flow where it really matters. Researchers are also looking for chemical markers in the blood or spinal column, which can help pinpoint dysfunctional components in the brain. And novel brain-scanning methods can measure neurotransmitter changes as a result of medication or mental illness.

The hope is that this multipronged attack will soon lead to a more tailored, neuroscience-based diagnosis to psychiatric syndromes, as well as a clearer idea of which existing drugs will help and which will not. And, ultimately, this should lead to better clues about the novel drugs we should be developing to target the root neurochemical causes of these problems.

BUILDING THE CONSCIOUSNESS MUSCLE

Clearly, the current suite of drugs used to treat many psychiatric conditions is not ideal, both in terms of efficacy and side effects. A far safer option, and one well worth trying, given the issues surrounding medication, would be to use behavioral treatment of some form, if it were effective. And if low working memory is a central cause of psychiatric symptoms in certain patient groups, such as those with ADHD or schizophrenia, then some therapy that targeted this problem may really help lift consciousness, giving these patients both the chance to process the world more appropriately and gain better control over their thoughts and behavior.

"Brain training," currently highly fashionable, might be one route to improving working memory. The kind of popular brain training that is played on game devices has recently, justifiably, received a bad press: Healthy nonelderly people don't gain any generalized improvement from practicing such games.*

But more clinical forms of brain training, designed specifically to boost working memory function, have shown much promise in alleviating mental illness.

For instance, Torkel Klingberg and colleagues gave ADHD children a battery of working memory tasks to practice for three weeks. They found the children experienced improvements not only in working memory and IQ levels but also in ADHD symptoms. Similar training in the scanner has been shown to boost prefrontal parietal network activity.

When pitted against medication, brain training yields even more interesting results: Joni Holmes and colleagues compared the ADHD working memory improvements due to Ritalin and due to working memory training. Quite surprisingly, there were far more dramatic and widespread working memory gains after training compared with medication—and these persisted for at least six months following the training regime. So for ADHD children, cognitive training may be a powerful route to boosting consciousness and control, all via working memory gains.

Amazingly, cognitive training even works on schizophrenia, the other psychiatric illness closely linked with profound working memory impairments. Melissa Fisher and colleagues used a commercially available cognitive-training program provided by Posit-Science, a San Francisco company, to apply intensive practice in working memory tasks, in this case for at least 50 hours. The schizophrenic patients improved both their cognitive skills, and their ability to function in day-to-day activities, with greater training improvements related to a better ability to cope with life. These improvements were also still present six months after training.

---

*There's some provisional evidence, however, that cognitive training is useful in staving off dementia in those over the age of sixty-five.

On the one hand, it seems quite surprising that simple working memory practice can alleviate some of the symptoms of these serious psychiatric illnesses. On the other hand, it makes sense that opening up a wider expanse of awareness would lead to an easier avoidance of false patterns and a host of improvements in control, information processing, and strategizing around problems.

A SEESAW OF STRESS VERSUS CONSCIOUSNESS

Although ADHD and schizophrenia are characterized by a lower-than-average working memory, certainly not all mental illnesses show this deficit as a central, continuous feature—even if a temporarily reduced conscious capacity in acute episodes does partially explain many conditions. In fact, a label of reduced mental attributes doesn't even apply to all schizophrenics: Although the majority of schizophrenic patients, before the illness takes hold, already have an impoverished intellect and low working memory, there seems to be a pocket of patients at the opposite end of the spectrum—those who are highly intelligent, and, for a time, tremendously productive.

One of the most famous examples is John Nash, a Nobel laureate in economics, whose story was brought to life by Sylvia Nasar's book *A Beautiful Mind*, which also became a Hollywood blockbuster movie. Just as Nash was trying to cement his academic career, his prodigious intellect started to unravel. Instead of continuing to use mathematics to unlock a wide range of subjects in a brilliant manner, his focus shifted to outlandish conspiracies; grandiose, utterly unattainable plans; and a host of paranoid delusions.

He started to spot bizarre patterns, coming to believe, for example, that anyone wearing a red tie was a member of a "crypto-communist party." He would write frantic notes to colleagues complaining that aliens from outer space were ruining his career. He also sent letters to all the embassies in Washington, D.C., declaring that he was forming a world government and wanted to discuss the matter with them. It was his plan to become emperor of the Antarctic. Although he continued to believe his work was of the highest caliber, the ideas were by this stage nonsensical, and soon the illness took full hold and his mind was ravaged by insanity.

During the time leading up to his first schizophrenic episode, Nash was mentally exhausted and probably chronically stressed from the combination of his teaching duties and his stupendously abstruse mathematical research. His wife was also about to give birth.

Nash's case raises the possibility that it is this very propensity to obsess over deeply challenging mental problems that contributes to mental imbalance in this small pocket of previously mentally superior individuals. Those like Nash are particularly skilled at searching for structure in the world and highly predisposed to devoting hours, days, and weeks to solid concentration on a single topic—which in itself probably places a great strain on their usually highly able prefrontal parietal network. Once the cancerous strands of this imbalance take hold, their conscious machinery can then, far more readily than most, become an internal enemy as it generates intricate paranoid delusional theories, which ramps up the stresses exponentially.

It's well known that stress is the single biggest trigger for almost all mental illnesses. As Nash's tragic life demonstrates, it can mangle the most able mind and plunge a person into years of nightmarish suffering. So in order to understand mental illness more generally, we need to understand how stress arises and what effects it has on our awareness and cortex.

If we're seriously under threat and our lives are in danger—for instance a car, out of control, is headed straight for us—then there is a tension about how best to respond. On the one hand, action is the key—we want to make a decision as fast as possible to move in order to get out of the way. On the other hand, we don't necessarily want to make too rash a decision to act, as we may run straight into the path of another car, or fail to realize that there's a concrete reinforced barrier we could hop over and protect ourselves with. This tension is reflected in our emotions and thoughts. First comes our primitive fear, which floods our body with chemicals in readiness for movement and orients us to the danger. The amygdala, our fear center, tends to suppress prefrontal cortex activity so that we don't think too much, and we act swiftly in whatever automatic way will keep us alive. On the other hand, the prefrontal parietal network can make a slower, more sophisticated, deliberative, conscious assessment of the situation—say, by noticing that although we appear to be in danger, the car moving straight for us is an illusion created by the movie projector on the cinema screen—and we are quite safe. So the half-strangled prefrontal cortex can, in turn, suppress amygdala activity, along with our feeling of fear, and bounce back to retain full conscious control of our movements.

In humans, in real life, the amygdala wins the battle all too often, and for entirely unnecessary reasons. If we were living, as our ancestors did, in a particularly threatening world, where big cats commonly stalked us, cantankerous rhinos and elephants regularly charged us, aggressive snakes liked to bite us, other members of the tribe sometimes wanted to fight us, and competing tribes sometimes declared war on us, then it arguably would be rather useful to be prone to surges of fear. However, it's not entirely clear that a chronically fearful life, even for our hominid ancestors, would have been beneficial, except in the most unusual, extreme times.

The simple fact is that, in the vast majority of cases, conscious understanding and innovation is a superior tool to instinctive fear. Every one of us faces potentially mortal threats every day. Walking down the stairs could kill us, as could crossing the road. But we learn over our developing years to give such events respect, and we maintain simple routines so that the events seem safe. This is in stark contrast to those animals with less intellect, such as cats, which probably have a burst of terror whenever they cross a busy road.

Most people, especially in the developed world, live safer lives than ever before. But we still carry within us this evolutionary legacy, this primitive machinery of fear, which all too easily latches onto any target we consciously deem important, regardless of whether it is a threat to our lives. So we may be intensely nervous when having to give a public speech, or when playing in a sports tournament. With our prefrontal parietal network shut down by our anxious amygdala, we might be keenly aware of the object of our fear, but the world outside that one meaningful object dims, and those tasks requiring consciousness, such as answering questions during a talk, suddenly seem almost impossible.

Thus one major factor in stress is the inappropriate labeling of a multitude of safe events as life threatening. This in itself arises out of a combination of the primacy of our primitive, unconscious drives in controlling our thoughts and behaviors, and the plethora of habits we form in life, partly in the service of such impulses.

Our propensity for chunking, for consciously seeking out those little habitual nuggets, in other ways so successful in giving us control of our environment, can fail catastrophically when applied to our inner emotional world. It's easy to underestimate the dizzying extent to which our lives revolve around the deep-seated chunks we've developed over the years. We are born utterly defenseless, with few instincts, but have minds exquisitely built for learning. Virtually every aspect of how we move, act, or carry out our daytime profession or evening hobbies is little more than a vast set of chunks we've learned over a lifetime. From the way we greet our parents on the phone to how we brush our teeth to the way we turn off our computer, every little detail of our lives involves the recalling of a piece of structured memory, a well-worn strategy, so that we have a fine-grained control over our world and an awareness that's free to notice and react to new or complex events.

Amid these myriad chunks, though, there will be some examples in all of us that cause us to react in an unhelpful, irrational way to a given event. For depressives like my wife, not getting absolute top marks on a test might trigger a set of thoughts and memories that will cause her to think she's utterly stupid, for instance. I can occasionally feel nervous talking to certain people not because of what they may say to me, or even what they ever did say, but simply because I once felt nervous in the past when talking to them, and this has become an inextricable habit. Because these are such well-used chunks, we are barely aware of them, just as we are barely aware of the mechanics of performing a forehand stroke in tennis.

This destructive form of chunking is particularly relevant to those with a mental illness. They might initially inappropriately label a given event as presenting an immense threat, as many of us do, but then fail to notice there was no threat after all. Instead, they may consciously chunk the link between the event and the level of danger so that it becomes reinforced and well established. Once this crystallizes in their minds, they will then repeatedly feel unavoidably, intensely threatened by these and any similar experiences. This can easily be compounded by further little chunks or tricks that are designed, in vain, to help. For instance, someone may be terrified of seeming awkward and foolish in social situations. But developing the habit to avoid all social moments means that this individual can never discover evidence to counteract his fears. Someone else may be terribly self-conscious of a small spot on her face, so she develops the habit to smother her whole face in thick makeup whenever she interacts with people, which means that she seems abnormal for this reason instead.

So while we are ingenious at understanding and innovation, we are unfortunately just as adept at forming thousands of personal strategies that can create chronic traps of stress and suffering, handing the advantage away from the prefrontal parietal network and to the amygdala. And when we are repeatedly exposed to stressful or frightening situations, then a damaging neural bias can develop: Our brains become entrained to overactivate the amygdala. This in turn can create a negative feedback loop that further inhibits activity in our prefrontal parietal network, reducing both our conscious space to deal with problems and our conscious control to filter out unhelpful thoughts and feelings.

The situation can then become semipermanent, with this imbalance continuing outside of episodes. For instance, if depressive or anxious patients are shown fearful stimuli in the scanner, then, understandably, they have greater amygdala activity than controls. But even if they are performing a neutral task requiring working memory or attention, these patients show reduced activity in parts of the prefrontal parietal network, as if these regions were chronically constricted, regardless of task.

Thus, a sustained period of stress or anxiety can largely turn off prefrontal function and consequently reduce conscious resources, with working memory half-disabled. In these situations, the brain learns by default to fear and to assume the most negative interpretation imaginable. The amygdala becomes trigger-happy, and its activity is sustained more aggressively, while the prefrontal parietal network is now easily and regularly inhibited. For schizophrenics, who may already have a far more fragile neurochemical system, stress may tip the scales toward catastrophic prefrontal dysfunction. But this mechanism may be an even more important clue to those mental illnesses more centrally defined by their emotional torments, such as post-traumatic stress disorder, anxiety disorder, and depression.

A WIDER, PURER OCEAN OF AWARENESS

If stress is such a profound game changer when it comes to mental illness, what can be done to reduce it and return the reins to the prefrontal parietal network? Aside from medication, which has only limited success, there is a practical list of lifestyle changes, which includes the usual suspects, such as avoiding or reevaluating those specific events that induce stress (not easy in most cases), getting decent quality sleep, exercising regularly, and so on. And these do all help reduce psychiatric symptoms, or have a protective effect against succumbing to a mental illness. But eclipsing this list in the fight against stress is one simple mental exercise: meditation. This is often written off as being too esoteric and not sufficiently scientific, but it's been shown to profoundly help virtually any mental ailment, whether the person has a psychiatric condition or is merely suffering from the stresses and strains of everyday life.

There needn't be anything mystical to meditation. Although there are hundreds of different varieties, in my view meditation is at its most powerful in its purest, most basic form: An ideal meditation is one where you try to be as aware as you can of as little as possible.

We can focus on the world with our attentional magnifying glass set in two broad modes: We can mainly attend to the thoughts, ideas, facts, and so on that relate to other mental events or what our senses are picking up; *or* we can simply attend to our senses directly, passively absorbing the experience without thoughts.

We may be on a mountaintop and feel a calm, profound awe as we attend entirely to the delicate beauty of the vast snowcapped peaks surrounding us. *Or* we could enter a very different experiential mode and try to calculate the volume of each mountain as a mathematical exercise, infer what the name of each mountain is from various clues, and recognize the different geological features of the rocks around us. In one state we have an open, quiet, passive perspective. In the other, we have a narrower, chattering mind, which isn't nearly so aware of our sensory experiences. This quieter sense of beauty is closely related to the so-called meditative experience.

Spending 20 minutes or more with nothing to do may sound pointless and tedious, but for those who try it, boredom, strangely, never seems an issue. It's as if, by intensely focusing attention continuously on the darkness of your closed eyes, a blank wall, or something equally minimal, you are telling your brain that this object is utterly fascinating. Early in meditation practice, there may be a struggle as the mind wanders, but after a while it becomes easier to sustain attention on virtually nothing for long periods of time.

And, far from being ineffectual, meditation causes substantial brain changes, both in the short term of a single session and in the long term, after months or years of practice. What's more, these brain alterations seem the exact reverse of those caused by stress and many mental illnesses.

In striking contrast to the effects of anxiety and stress, the simple act of entering a meditative state increases activity in the prefrontal parietal network, especially in the lateral prefrontal cortex. So, intriguingly, this is indirect objective evidence that meditation really does raise awareness.

Over years of practice, regular meditation also seems to permanently change the prefrontal parietal network, such that it becomes more pliable and efficient in its activity. And, again in direct contrast to what happens in depression and anxiety disorders, long-term meditation shifts the see-saw battles between the amygdala and prefrontal cortex firmly in the prefrontal cortex's favor: The amygdala becomes far less likely to activate, probably partly because the prefrontal parietal network is now so good at stepping in and taking control. There is even evidence that long-term meditation increases the thickness of the prefrontal cortex, helping to protect against the natural thinning of this part of the brain in old age. At the same time, two months of meditation is sufficient to shrink the size of the fear-creating amygdala in previously stressed individuals.

All these results fit perfectly with long-term meditators' descriptions of how this practice modifies their experiences. They report becoming profoundly calm, largely free from fear, and better able to handle pains and bothersome emotions when they do arise. They describe a greater degree of awareness and mental control, as if they somehow have more space and flexibility by which to perceive and handle all the details of the world and their own inner life. It's as if, by establishing new attentional habits that shift focus away from highly grooved mental chunks, any unhelpful old ideas can more easily be displaced. When it comes to our emotional habits particularly, meditation is undoubtedly an invaluable tool to dislodge those painful schemas that might have housed themselves firmly in our unconscious minds, and which can be the source of so much of our painful thoughts and feelings.

If meditation really can cause awareness to expand in a nourishing way, and consciousness is intimately linked with attention and working memory, then meditation should in turn generate improvements in tasks tapping these processes. This is exactly what researchers are discovering: Long-term meditation does improve a range of attentional tasks as well as working memory skills and spatial processing. Strikingly, regular meditation also seems to reduce a person's need for sleep, possibly because it is a neurally nourishing activity.

However, one needn't spend many years in intensive meditation practice before any benefits are seen. Fadel Zeidan and colleagues, for instance, found that just four meditation sessions were sufficient to reduce feelings of tiredness and increase working memory performance. Another study, by Yi-Yuan Tang and colleagues, found that only five days were needed for volunteers to improve on an attentional task that measured the subject's ability to deal with conflicting stimuli. In addition to this improvement in a central component of cognition, the volunteers felt less anxiety, depression, anger, and tiredness.

Therefore, for normal, healthy participants, meditation reduces stress, improves alertness, and enhances performance on an impressive array of demanding tasks. It's unsurprising, therefore, that meditation is increasingly being used as an effective weapon against depression, anxiety disorder, severe pain management, schizophrenia, and a host of other conditions.

HEALING CONSCIOUSNESS FROM MANY ANGLES

Along with our astounding conscious capacity to understand the universe and fashion increasingly sophisticated tools to control it, we have uniquely rich experiences punctuating our lives. The price for all this, though, is sharper suffering. Some traumas are to be expected, given that we lead such extensive, long lives, and thus have many decades by which to experience the sorrows as well as the joys. But we are also prone to, by far, the widest, most intense range of mental illnesses of any species, and these can cripple us as commonly as any disease of the heart or lungs. These heavy costs of a vast, immensely capable consciousness are themselves intimately related to awareness at every turn. While autism, the odd-one-out, partially relates to an overabundant awareness, other conditions reflect a permanent or episode-driven diminished consciousness, with working memory reduced and attention failing to filter out unhelpful thoughts and feelings. This combination creates an attachment to the aberrant and upsetting structures of thought that so heavily reinforce mental diseases.

Research is revealing many ways to soften symptoms and approach cures, with each method presenting a different angle to the challenge of returning consciousness to normal levels. Current and emerging drugs can rebalance a dysfunctional neurotransmitter system and in turn return prefrontal parietal processes to their optimal levels. Future tools of psychiatric diagnosis, however, need to give more emphasis to neuroscientific dysfunction, and appropriate drugs need to be provided that are tailored to meet this specific dysfunction. Cognitive training and meditation offer complementary routes to boosting working memory and refocusing attention. And all these methods can encourage a new sense of control over one's inner mental world.

All the time, the details of mental-illness symptoms and the array of useful treatments reaffirms the view of consciousness as involving the prefrontal parietal network to support a combination of attentional and working memory systems, with pattern-searching a crucial function.

Although some of these nascent treatments haven't as yet reached the clinic, hopefully this soon will change. And by holding in mind the putative perspective that each of these mental illnesses intimately relates to a skewed, misfiring consciousness, clinical researchers may make important new breakthroughs. It's vital that this army of future strategies be given as much attention and as many resources as possible so that the pandemic of mental illness can be overcome.

Epilogue

A Delicious Life

The science of consciousness is coming of age. It can explain the origin and purpose of awareness, its mental features and neural mechanisms, as well as its intense fragility. From this picture, we can see the seeds of awareness in the history of life on earth. All ancient creatures blindly clung to survival by capturing and combining useful ideas about the environment—but this limited learning happened randomly, awkwardly, via DNA changes and the rhythmic fall of generations.

Consciousness first arose out of this evolutionary endeavor via a specialist information-processing organ, a neural computer capable of acquiring accurate concepts and deep strategies far more rapidly and flexibly than ever before. Some evolutionary branches created complex brains, which were optimized to build pyramids of knowledge out of simpler ideas cemented together. This critical skill in combining mental objects to generate increasingly meaningful, useful structures of thought is both the essence of consciousness and its overriding purpose.

Humans have a unique place in the world. We have an exceptionally complex brain, whose central wired core, in the form of the prefrontal parietal network, is greatly expanded even compared to our nearest primate relatives. We can process and combine information like no other species, and consequently we experience the world in exceptionally rich, varied ways.

Our consciousness allows us to unlock nature's secrets, from the architecture of the smallest atom inside us to the swirling mist of galactic stars above us. Our consciousness enables us to ward off death by decades via medical technology, to travel to the moon in a couple of days, and to generate a plethora of immensely sophisticated gadgets by which to feed our prodigious appetite for stimulation. We should feel incredibly fortunate that evolution has endowed each of us with this immense biological computer, which can experience and control the world with such variety and depth.

But our pride in the mental richness of our lives and the power of our intellect should be tempered by a nagging sense that, in the case of human brains, evolution almost seems to have overreached. The signature skill of our conscious minds, our advanced powers to search for profound patterns that help us understand or conquer our environs, was only ever meant to serve our primitive evolutionary drives of survival and reproduction. This observation is most apparent in the vertiginous chasm that exists between the enormous wealth of conscious comprehension that any educated adult demonstrates daily and our embarrassing ignorance of the reasoning behind so many of our decisions. Some of us strive for a more enlightened life, somewhat divorced from these basic impulses, but such undercurrent motivations nevertheless rule and constrict us. Only in humans do these forceful evolutionary drives seem sometimes to clash so violently with higher conscious goals.

Moreover, we are so clever at spotting the patterns and tricks to meet our primitive desires that our lives can easily spiral out of control. Humanity's prodigious conscious tools of innovation can be devoted to discovering broad truths about the world. But they are just as easily co-opted to generate inventive tactics to have affairs, to overeat, to steal, or to pursue all manner of other short-term goals that are likely to backfire when all the consequences are counted up.

And, all too easily, our aggressive ability to form such packets of ideas and behavior perpetuates unhappy traps of thought, or even, occasionally, outright delusions, as we unwittingly reveal the delicate fragility of the human mind. For some, this route leads to a debilitating mental illness, with consciousness mutating into an enemy—to such an extent that a few are so tormented that they wish to end their lives, to free themselves of this toxic inner foe. But all of us, to varying degrees, are both the beneficiaries and the victims of our own consciousness.

I'm utterly passionate about science, not just because of the wondrous, surprising, fascinating facts it reveals about the universe and our place within it but also, more abstractly, for its dogged devotion to the truth. As a romantic at heart, I like to think there is nothing more noble than this meticulous, ruthlessly honest, strangely humble goal to understand the intricacies of reality. But as part of my affection for science, I revel as well in the pragmatic reach of this professional investigation into natural machinery: Scientific discoveries effortlessly engender a multitude of powerful, malleable, liberating tools of technology. Evolution has learned the lesson a billion times over that rational, accurate ideas about your local world, whether they are stored in DNA, proteins, or brains, equates to control and success, even in the teeth of the most severe habitat. But humans, exceptionally, have an attentional system that can label absolutely anything as potentially biologically relevant and push it forward for deeper conscious analysis. Science is the crystallization of this widely roaming curiosity, and its technological fruits pervade modern life.

And, as with so many other branches of science, a rapidly maturing scientific picture of consciousness should yield perspectives and technologies that improve our lives. Although we have developed methods to communicate with brain-damaged patients by their thoughts alone, and can pharmacologically target brain chemicals and regions to return consciousness to normal levels in some mentally ill patients, ideally we'd expect such an intimate scientific field to help every one of us.

Already, the emerging scientific revelations about awareness can help us tip the scales away from being victims of our own consciousness and toward relishing the breathtaking range of skills and experiences awareness can provide. First, there are the rather trivial issues, such as that we should be wary of stress, and shouldn't underestimate the importance of sleep, both for its protective role against mental illness and its ability to keep our consciousness as clear and wide as possible.

In addition, certain more abstract scientific results, some deeply unintuitive, can help soften our view of ourselves and smooth our interaction with others. Our consciousness is by default the loyal clever assistant of our dumb primitive unconscious, which is in many instances running the show. Because of this, at times our thoughts and decisions may be far from smart. But this imbalance of power and limited control is the product of evolution's addiction to selfish, demanding traits; every single one of my billions of ancestors, stretching back to the first life-forms, had to possess such ruthlessly self-interested drives in order for me to exist today. This doesn't necessarily negate our responsibility for our actions, but it can make us more tolerant of our foibles and mistakes. And as we more readily recognize when our unconscious impulses are blindly guiding us, we have more chance to change tack, if necessary.

We can also apply the same sympathy to others, who may be far less able to curb their actions than they realize. Given the pandemic of mental illness, many may secretly be harboring depression or another such disease, and any apparently hurtful behavior they exhibit could merely be a result of a complete lack of control and a skewed awareness. This scientifically driven "benefit of the doubt" can help us view others' actions with more tolerance and acceptance, and less anger.

But there are far deeper ways that the science of consciousness can help us change the way we live. Recently, I watched my baby daughter learn to walk. Walking is one of those activities that adults take for granted. But for my infant daughter, absolutely everything in life is a wonderful toy, and walking is included in that inexhaustible list, like everything else. As soon as she mastered this new skill, she passionately ambled around the house almost every waking moment, to experience the unbridled fun of rhythmically plodding her feet shakily in front of her. Sometimes she walked sideways, or in circles; occasionally she even indulged in a good bit of stamping. She was especially proud and excited when she managed to walk backward. Walking barefoot, with socks, and, on special occasions, with shoes each provided their own species of gleeful exploration. She even found it particularly funny, unfortunately for me, to walk across my stomach and chest if I was lying down. Toddling around seemed to provide an infinite variety of entertainment for her.

To my baby daughter, everything is exciting, because she is filling her fledgling awareness every day with new ways of seeing and understanding the world. She has a passionate, open readiness to form novel chunks. I find this approach utterly beguiling and infectious, but at the same time, can't help comparing it to my own more closed view of life.

As we all emerge into adulthood and slowly age, there is a natural transition away from this playful building of mental chunks. We gradually replace this perspective with a more measured, perhaps even dulled approach, as we're weighed down by the myriad chunks we've already built up over our lifetime. In some ways it makes perfect sense that we increasingly turn into creatures of such extensive habit, because we've already learned many vital chunks of information, many ready tricks for interacting with the world. Successful bacteria in stable environments take the safer, less innovative route with lower mutation rates. And we do the same: As we get older, our learning increasingly protects us, so why rock the boat by fiddling with the formula now?

The trouble is, though, that the initially dazzling glow of our experiences in infancy, when our hunger for patterns is particularly urgent, can be dimmed and obscured over the years by so many overly familiar nuggets of knowledge. We are less ravenous for new jewels of wisdom, and our entire existence, examined through the perspective of the thousands of chunks we've acquired, can become routine.

What we need is a way to crank up our conscious levels in a more immersive way. After all, the more conscious we are, the brighter, more vibrant, and more pregnant with opportunities the world appears. This is trivially true in experiments in the lab: When I focus attention on a location on the computer monitor, it makes the upcoming dot stand out more when it appears there. But it's profoundly, wonderfully true in real life, when I now shine every ounce of my beam of attention on the broad smile on my baby daughter's face.

How does the science of consciousness help here? Well, if consciousness is all about innovation and shunting automatic habits to our unconscious, we can try to foster more awareness by biasing our minds as far as possible in the direction of innovation and relying at least a little bit less on our bank of deeply grooved habits.

One potentially effective strategy along these lines to help combat our slow, age-related deterioration of conscious fire is to nurture a globally questioning, doubting perspective. If consciousness is fundamentally about being ravenous for those innovations that will help us in life, then by fostering a habit of skepticism of almost every mental chunk that passes through our consciousness, we're setting up a superchunk, as it were, a higher habit to be restlessly searching for any little innovation we can get our hands on.

It's generally not so hard to detect and question those basic drives that try to steer our thoughts and behavior. But far more slippery, pervasive, and controlling are the thousands of ready-made mental packets, once consciously and carefully pieced together, that are now so automatic that we hardly notice them. Most of these chunks are invaluable in our lives. For instance, I give little attention to how I am touch-typing these words on my computer keyboard, and so have more conscious space to write my sequence of ideas. But a fair few chunks, just as automatic and long since submerged below my awareness, involve how I deal with my emotions, my relationships with others, and various other fields affecting my well-being. Some of these may be habitually generating fears, eroding my self-esteem, robbing me of happier moments, or polluting intimacy with my family and friends.

Again, though, the science of consciousness can help us dislodge and nullify these toxic patterns of behavior. It immediately helps to acknowledge that we're little more than a mental bag of such mini-programs, and that all of them can be revised, rewritten, or even nullified by other chunks we establish in their stead.

More importantly, we can spend considerable time trying specifically to notice what old structured chunks invade our emotions, decisions, and behavior, especially if they seem to underlie feelings of uneasiness. We can bring them briefly back to consciousness, examine their shape and source, and ask ourselves whether they are helping us or need to be modified.

In fact, because we are all so quick to spot patterns, even specious ones, it helps more generally to acknowledge that we are in some ways a little too hungry for knowledge—that all manner of spurious beliefs are regularly going to take anchor, so we need to be on our guard with a cautious, distanced approach to any ideas we entertain, in a way that mirrors the obligatory scientific skepticism in the lab.

Simultaneously, we can constantly nurture a searching attitude for useful and exciting new chunks to absorb.

All this questioning of preexisting chunks and gathering of fresh ones to supplement or supplant them may sound like unsettling, needless extra work, but it can be surprisingly pleasurable and invigorating to be perpetually skeptical about our own ideas and beliefs, partly because it means we're constantly on the lookout for new ways to live. It makes us feel somehow newer ourselves, less stale, more dynamic, more open to new experiences—even more alive.

The practice of meditation very much complements these questioning habits.

Meditation is increasingly proving to be a powerful tool to boost and calm consciousness in mentally ill patients. But it can also help us all experience the world in new, more connected ways. By adopting an immovable, piercing, open attention directed only toward the contents of our sensations, along with an awareness temporarily unfettered by thoughts, we can choose, for a time, completely to reject the plethora of strategies and habits we've built up over the years. Instead we enter a welcoming, ready state where we're deliberately labeling every speck of information flowing into us as important and unexpected.

A meditative mind can be strangely reminiscent of how we experienced the world as a child. We're exquisitely ready for fresh illumination, hungry and wide open for novel insights, but at the same time, deliciously bathed in the present moment. The rich, immediate diversity of direct sights, sounds, and smells comes once again to the fore. Without the myriad mental obstacles of those chunks invading our thoughts, we can reacquaint ourselves with how beautiful so much of the world really is, and how very easy it is to find intense pleasure and joy within it.

Previously, buried under our usual mass of habits, we might have ignored the taste of a whole plate of dinner as we simultaneously watched the latest TV drama. But now, simply, silently eating a meal is an overwhelming treat of stimulation: We devote every ounce of awareness to every bright facet of flavor of every single morsel—and it never tasted so delicious.

Acknowledgments

Throughout the process of writing *The Ravenous Brain*, I've continuously felt both terribly lucky and immensely grateful to receive invaluable help from such a large number of people, and I shudder to think how the book would have turned out without them.

At the early stages, my friend and former work colleague John Duncan demonstrated his kindness and generosity yet again with critical assistance in shaping my proposal. My fabulous agent, Peter Tallack, has helped in so many ways, initially with hard work on editing the proposal, then with securing the deal; later with looking over various chapters, not to mention those friendly lunches, and consistently quick responses to my bothersome stream of questions about the process.

Continuing the theme of luck, I managed somehow to receive not one very talented editor, but two. TJ Kelleher's very wise advice at multiple times both on the overall shaping of the book and on the content of specific sections both shocked me for the extent of flaws in the manuscripts I sent him and excited me for the opportunities he showed me for dramatic improvement. Tisse Takagi has helped enormously in cutting out the flab from the chapters, reducing my vagaries, and smoothing out my stylistic foibles, not to mention being another person to gracefully and swiftly respond to my bothersome stream of questions. My copy editor, Katherine Streckfus, did a wonderful job of correcting all those unclear or stylistically awkward lines. Then, in the final stages, my production editor, Michelle Welsh-Horst, was an exceptionally efficient, friendly manager as she guided the book from being a rough manuscript to the product you have in your hands.

Many family members also lent a hand. My cousins Adam Beckman, Fiona Beckman, and Eleanor Baker, and parents-in-law Ramarao Paidisetty and Indira Patra, put me on the right track, especially during the early stages. My cousin, Janetta Lewin, gave invaluable advice on some of the artistic issues. My uncle, Tony Epstein, provided very useful financial advice. My mother, Gaynor Bor, and brother, Simon Bor, tirelessly read through multiple drafts of each chapter, pointing out unclear sections and other problems.

Friends and colleagues at Cambridge's Medical Research Council (MRC) Cognition and Brain Sciences Unit who gave great advice and encouragement on various chapters include Adrian Owen, Jessica Grahn, Martin Monti, Rhodri Cusack, and Lorina Naci, while Simon Strangeways performed his photographic magic to create my cover portrait. I received further help from many at my current department, the Sackler Centre for Consciousness Science at the University of Sussex. In particular, Anil Seth read through an entire late draft and made many excellent technical comments, while advice and discussions from Ryan Scott, Adam Barrett, and others in the center helped me improve various tricky passages.

Further afield, Bernard Baars helpfully and graciously offered his knowledge, while Jonathan Huntley had many insightful suggestions for the psychiatry chapter. In addition, there were so many others in the research community over the years who have imparted their wisdom during chats over coffee or a beer, but this list would take too long to write here, so apologies for not naming you explicitly.

Finally I owe an immense amount of gratitude to the two people closest to me. My daughter, Lalana, while not *quite* at the stage of offering me advice, unflinchingly produced an excited smile and open arms on my return home, despite guilty periods when I worked so hard that I would hardly ever see her. And she's been a wonderful inspiration for this book, as I was able to see her consciousness develop as the manuscript did. Greatest thanks of all, though, goes to my wonderful wife, Rachana, for taking up the household slack fantastically during my obsessive writing phases (no mean feat when you have an insomniac baby around!), for reading every draft of every chapter (and there were *a lot* of drafts!), making copious observant comments each time, and for generally being so supportive throughout.

Notes and References

Chapter 1: Conceptual Conundrums of Consciousness

4   **In his most famous work, *Meditations on First Philosophy***

R. Descartes, J. Cottingham, and B.A.W. Williams, *Meditations on first philosophy: with selections from the objections and replies.* 1996, Cambridge: Cambridge University Press.

5   **"*Cogito ergo sum*"**: R. Descartes and I. Maclean, *A discourse on the method.* 2006, Oxford: Oxford University Press.

8   **A landmark paper . . . described . . . Gage**

J. M. Harlow, Recovery from the passage of an iron bar through the head. *Publ Mass Med Soc* (Boston), 1868. 2: 327–346.

8   **Controversy about the behavioral details**

M. Macmillan, *An odd kind of fame: stories of Phineas Gage.* 2002, Cambridge: MIT Press.

8   **Seminal 1949 work, *The Concept of Mind***

G. Ryle and D. C. Dennett, *The concept of mind.* 2002, Chicago: University of Chicago Press.

10   **But more as a *superorganism***

B. Hölldobler and E. O. Wilson, *The superorganism: The beauty, elegance and strangeness of insect societies.* 2009, New York: Norton.

11   **What Is It Like to Be a Bat?**

T. Nagel, What is it like to be a bat? *Philos Rev*, 1974. 83(4): 435–450.

12   **Impossible . . . explain mental states using only physical processes**

F. Jackson, Epiphenomenal qualia. *Philos Q*, 1982. 32: 127–136.

F. Jackson, What Mary didn't know. *J Philos*, 1986. 83(5): 291–295.

13 **Not as watertight as it might at first appear**

D. C. Dennett, *Consciousness explained.* 1991, New York: Penguin.

14 **Jackson . . . has since rejected his former argument**

F. C. Jackson, Mind and illusion, in *Minds and persons,* A. O'Hear, ed. 2003, Cambridge: Cambridge University Press.

16 **Swap red with blue . . . leave all thought . . . unchanged**

A. Byrne, Inverted qualia. 2010; http://plato.stanford.edu/entries/qualia -inverted/.

S. Shoemaker, Functionalism and qualia, in *Readings in the philosophy of psychology,* N. Block, ed. 1980, Cambridge: Harvard University Press, 251–267.

16 **No single, independent class of experience as "red"**

Dennett (1991), see above.

17 **John Searle in 1980**

J. R. Searle, Minds, brains and programs. *Behav Brain Sci,* 1980. 3(3): 417–424.

23 **85 billion neurons in a human brain**

F. A. Azevedo et al., Equal numbers of neuronal and nonneuronal cells make the human brain an isometrically scaled-up primate brain. *J Comp Neurol,* 2009. 513(5): 532–541.

24 **Micro-cables . . . wrap around the earth four times**

L. Marner et al., Marked loss of myelinated nerve fibers in the human brain with age. *J Comp Neurol,* 2003. 462(2): 144–152.

25 **Computer models closely approximating . . . population of neurons**

E. M. Izhikevich and G. M. Edelman, Large-scale model of mammalian thalamocortical systems. *Proc Natl Acad Sci USA,* 2008. 105(9): 3593–3598.

28 **Tatiana and Krista**

S. Dominus, Inseparable. *New York Times,* May 29, 2011, MM28.

CHAPTER 2: A BRIEF HISTORY OF THE BRAIN

36 **Life is honed from natural experimentation**

M. Gell-Mann, *The quark and the jaguar: adventures in the simple and the complex.* 1994, New York: W. H. Freeman.

38 **Self-assembling non-life molecules . . . technologically exploited**

J. R. Nitschke, Systems chemistry: molecular networks come of age. *Nature,* 2009. 462(7274): 736–738.

41 The default state for networks of neurons

J. M. Beggs, The criticality hypothesis: how local cortical networks might optimize information processing. *Philos Transact A Math Phys Eng Sci,* 2008. 366(1864): 329–343.

45 Creating . . . bacteria to make diesel fuel

A. Schirmer et al., Microbial biosynthesis of alkanes. *Science,* 2010. 329(5991): 559–562.

46 Human genes introduced into mice

W. Enard et al., A humanized version of Foxp2 affects cortico-basal ganglia circuits in mice. *Cell,* 2009. 137(5): 961–971.

46 Mouse genes into flies

G. Halder, P. Callaerts, and W. Gehring, Induction of ectopic eyes by targeted expression of the eyeless gene in Drosophila. *Science,* 1995. 267(5205): 1788–1792.

46 Plant species . . . formed . . . of two or more . . . lineages

L. H. Rieseberg, T. E. Wood, and E. J. Baack, The nature of plant species. *Nature,* 2006. 440(7083): 524–527.

46 Gene swaps between humans and viruses or bacteria

S. L. Salzberg et al., Microbial genes in the human genome: lateral transfer or gene loss? *Science,* 2001. 292(5523): 1903–1906.

S. Mi et al., Syncytin is a captive retroviral envelope protein involved in human placental morphogenesis. *Nature,* 2000. 403(6771): 785–789.

49 Mutation rates are increased in some bacteria

I. Bjedov et al., Stress-induced mutagenesis in bacteria. *Science,* 2003. 300(5624): 1404–1409.

49 Yeast . . . reshuffling . . . entire chromosomes

G. Chen et al., Hsp90 stress potentiates rapid cellular adaptation through induction of aneuploidy. *Nature,* 2012. 482(7384): 246–50.

49 Primates . . . lowest social standing . . . exhibit innovative behaviors

S. M. Reader, Primate innovation: sex, age and social rank differences. *Int J Primatol,* 2001. 22(5): 787–805.

51 Faced with some threat . . . worms . . . forgo . . . self-fertilization

L. T. Morran, M. D. Parmenter, and P. C. Phillips, Mutation load and rapid adaptation favour outcrossing over self-fertilization. *Nature,* 2009.

51   **All organisms . . . lives extended . . . simply by eating less**

S. D. Hursting et al., Calorie restriction, aging, and cancer prevention: mechanisms of action and applicability to humans. *Annu Rev Med*, 2003. 54: 131–152.

52   **Near perfect memory . . . of Solomon Sherashevski**

A. R. Luria, *The mind of a mnemonist.* 1966, Cambridge: Harvard University Press.

56   **Greatest source of evolutionary innovation is the virus**

F. Ryan, I, virus: why you're only half human. *New Scientist*, 2010. 2745: 32–35.

56   **50 percent of our genome . . . from ancient viruses**

E. S. Lander et al., Initial sequencing and analysis of the human genome. *Nature*, 2001. 409(6822): 860–921.

56   **Viral donation of DNA . . . placenta in early mammals**

Mi et al (2000), see above.

57   **Dawkins' language they are termed "survival machines"**

R. Dawkins, *The selfish gene: 30th anniversary edition.* 2006, Oxford: Oxford University Press.

R. Dawkins, *The extended phenotype: the long reach of the gene,* 2nd ed. 1999, Oxford: Oxford University Press.

57   **Million copies of the Alu sequence**

Lander et al (2001), see above.

58   **Selfish imposition of a single mutation-causing gene**

Dawkins (2006), see above.

59   **20,000 genes . . . required to create your brain**

A. R. Jones, C. C. Overly, and S. M. Sunkin, The Allen brain atlas: 5 years and beyond. *Nat Rev Neurosci*, 2009. 10(11): 821–828.

59   **Bacteria . . . resemble a multicellular organism**

J. A. Shapiro, Thinking about bacterial populations as multicellular organisms. *Annu Rev Microbiol*, 1998. 52: 81–104.

60   **Douglas firs . . . share soil resources with saplings**

F. P. Teste et al., Access to mycorrhizal networks and roots of trees: importance for seedling survival and resource transfer. *Ecology*, 2009. 90(10): 2808–2822.

60 **Tomato . . . release . . . chemicals that neighboring plants can read**

Y. Y. Song et al., Interplant communication of tomato plants through underground common mycorrhizal networks. PLoS One, 2010. 5(10): e13324.

60 **Ecosystems . . . self-organize**

S. A. Levin, Ecosystems and the biosphere as complex adaptive systems. *Ecosystems*, 1998. 1(5): 431–436.

M. Rietkerk and J. van de Koppel, Regular pattern formation in real ecosystems. *Trends Ecol Evol*, 2008. 23(3): 169–175.

61 **Cell . . . computation**

D. Bray, *Wetware: a computer in every living cell.* 2009, New Haven, CT: Yale University Press.

63 **Protozoa and bacteria . . . learning and memory**

A. Mitchell et al., Adaptive prediction of environmental changes by microorganisms. *Nature*, 2009. 460(7252): 220–224.

63 **Limiting factors to this process**

S. B. Carroll, Chance and necessity: the evolution of morphological complexity and diversity. *Nature*, 2001. 409(6823): 1102–1109.

68 **Rewire the ferret visual pathway**

J. Sharma, A. Angelucci, and M. Sur, Induction of visual orientation modules in auditory cortex. *Nature*, 2000. 404(6780): 841–847.

L. Von Melchner, S. L. Pallas, and M. Sur, Visual behaviour mediated by retinal projections directed to the auditory pathway. *Nature*, 2000. 404(6780): 871–876.

68 **When reading Braille, process . . . in the visual regions**

H. Burton et al., Adaptive changes in early and late blind: a fMRI study of Braille reading. *J Neurophysiol*, 2002. 87(1): 589–607.

68 **Unconscious statistical machine**

C. Frith, *Making up the mind: how the brain creates our mental world.* 2007, Oxford, UK: Wiley-Blackwell.

70 ***C. elegans . . . can learn . . .***

M. De Bono and A. V. Maricq, Neuronal substrates of complex behaviors in *C. elegans. Annu Rev Neurosci*, 2005. 28: 451–501.

71 **Recognize our own emotions . . . our body states**

A. R. Damasio, *Descartes' error: emotion, reason, and the human brain.* 1994, New York: HarperCollins.

74 **Patterns . . . attractive to bees**

M. Lehrer et al., Shape vision in bees: innate preference for flower-like patterns. *Philos Trans: Biol Sci*, 1995. 347(1320): 123–137.

CHAPTER 3: THE TIP OF THE ICEBERG

82 **Three evolutionary versions of brains**

P. D. Maclean, *The triune brain in evolution: role in paleocerebral functions.* 1990, New York: Springer.

83 *The Diving Bell and the Butterfly*

J.-D. Bauby, *The diving bell and the butterfly.* 2008, New York: Harper Perennial.

84 **Parietal lobes . . . linked with . . . IQ**

J. R. Gray, C. F. Chabris, and T. S. Braver, Neural mechanisms of general fluid intelligence. *Nat Neurosci*, 2003. 6(3): 316–322.

85 **Frontal lobes . . . complex and novel**
Ibid.

J. Duncan and A. M. Owen, Common regions of the human frontal lobe recruited by diverse cognitive demands. *Trends Neurosci*, 2000. 23(10): 475–483.

J. Duncan et al., A neural basis for general intelligence. *Science*, 2000. 289(5478): 457–460.

E. K. Miller and J. D. Cohen, An integrative theory of prefrontal cortex function. *Annu Rev Neurosci*, 2001. 24: 167–202.

91 **When under . . . anesthesia . . . learning is beyond us**

J. F. Kihlstrom and R. C. Cork, Consciousness and anaesthesia, in *The Blackwell companion to consciousness*, M. Velmans and S. Schneider, eds. 2007, Oxford, UK: Blackwell, 628–639.

91 **One of Dijksterhuis's experiments**

A. Dijksterhuis et al., On making the right choice: the deliberation-without-attention effect. *Science*, 2006. 311(5763): 1005–1007.

92 **Delegate thinking . . . to the unconscious**
Ibid.

92 **Newspaper articles written all over the world**

A. Jha, Want to make a complicated decision? Just stop thinking. *The Guardian*, February 16, 2006, 11.

B. Carey, The unconscious mind: a great decision maker. *New York Times*, February 21, 2006.

BBC News, Sleep on it, decision-makers told. 2006; http://news.bbc.co.uk/1/hi/health/4723216.stm.

92 **Malcolm Gladwell**

M. Gladwell, *Blink: the power of thinking without thinking*. 2005, London: Allen Lane.

93 **Made up their minds before . . . seen all the facts**

L. Waroquier et al., Is it better to think unconsciously or to trust your first impression? A reassessment of unconscious thought theory. *Soc Psychol and Personality Sci*, 2010. 1(2): 111–118.

94 **No advantage for unconscious processing**

F. Acker, New findings on unconscious versus conscious thought in decision making: additional empirical data and meta-analysis. *Judgment and Decision Making*, 2008. 3(4): 292–303.

94 **Conscious deliberation provided an advantage over distraction**

B. Aczel et al., Unconscious intuition or conscious analysis? Critical questions for the deliberation-without-attention paradigm. *Judgment and Decision Making*, 2011. 6(4): 351–358.

B. R. Newell, K. Y. Wong, and J.C.H. Cheung, Think, blink or sleep on it? The impact of modes of thought on complex decision making. *Q J Exp Psychol*, 2009. 62(4): 707–732.

95 **Transfer that learning . . . while still believing . . . guessing randomly**

R. B. Scott and Z. Dienes, Knowledge applied to new domains: the unconscious succeeds where the conscious fails. *Conscious Cogn*, 2010. 19(1): 391–398.

96 **Distracted . . . then we learn absolutely nothing**

D. Tanaka et al., Role of selective attention in artificial grammar learning. *Psychon Bull Rev*, 2008. 15(6): 1154–1159.

97 **The unconscious mind is unable to cope**

R. F. Baumeister and E. J. Masicampo, Conscious thought is for facilitating social and cultural interactions: how mental simulations serve the animal-culture interface. *Psychol Rev*, 2010. 117(3): 945–971.

98 **Damage to the lateral prefrontal . . . don't generate . . . wrong theories**

G. Wolford, M. B. Miller, and M. Gazzaniga, The left hemisphere's role in hypothesis formation. *J Neurosci*, 2000. 20(6): RC64.

99 **No evidence . . . subliminal messages influence our behavior**

P. M. Merickle, Subliminal perception, in *Encyclopedia of psychology*, A. E. Kazdin, ed. 2000, New York: Oxford University Press, 497–499.

101 **Held . . . position . . . closeness of their relative**

E. A. Madsen et al., Kinship and altruism: a cross-cultural experimental study. *Br J Psychol*, 2007. 98 (Pt 2): 339–359.

101 **Event . . . placed Kahneman on the path of psychology**

D. Kahneman, Daniel Kahneman: Autobiography (from the Nobel Prize site). 2002; http://nobelprize.org/nobel_prizes/economics/laureates/2002/kahneman-autobio.html.

102 **Unconscious anchoring of our choice**

A. Tversky and D. Kahneman, Judgment under uncertainty: heuristics and biases. *Science*, 1974. 185(4157): 1124–1131.

103 **Benjamin Libet demonstrated this fact**

B. Libet et al., Time of conscious intention to act in relation to onset of cerebral activity (readiness-potential): the unconscious initiation of a freely voluntary act. *Brain*, 1983. 106 (Pt 3): 623–642.

103 **Similar unconscious source . . . decision to veto**

P. Haggard, Human volition: towards a neuroscience of will. *Nat Rev Neurosci*, 2008. 9(12): 934–946.

W. P. Banks and S. Pockett, Benjamin Libet's work on the neuroscience of free will, in *The Blackwell companion to consciousness*, M. Velmans and S. Schneider, eds. 2007, Oxford, UK: Blackwell, 657–670.

104 **Consciousness was smeared across time**

D. C. Dennett, *Consciousness explained*. 1991, New York: Penguin.

104 **Computational model by Stanislav Nikolov**

S. Nikolov, D. A. Rahnev, and H. C. Lau, Probabilistic model of onset detection explains previous puzzling findings in human time perception. *Front Psychol*, 2010. 1: 37.

105 **Detected up to 10 seconds prior to the conscious decision**

C. S. Soon et al., Unconscious determinants of free decisions in the human brain. *Nat Neurosci*, 2008. 11(5): 543–545.

CHAPTER 4: PAY ATTENTION TO THAT PATTERN!

111 **". . . Mummy sent me to fetch you"**

I. Stewart, *Professor Stewart's hoard of mathematical treasures*. 2009, London: Profile Books.

113 Brain . . . half of all the energy the child consumes

M. A. Holliday, Body composition and energy needs during growth, in *Human growth: a comprehensive treatise*, F. Falkner and J. M. Tanner, eds. 1986, New York: Plenum, 101–117.

113 Human brain . . . nearing the endpoint . . . biologically possible

D. Fox, The limits of intelligence. *Sci Am*, 2011. 305(1): 36–43.

114 Two photos identical except for one feature

R. A. Rensink, J. K. O'Regan, and J. J. Clark, To see or not to see: the need for attention to perceive changes in scenes. *Psychol Sci*, 1997. 8(5): 368–373.

115 Half the volunteers notice . . . person has changes

D. J. Simons and D. T. Levin, Failure to detect changes to people during a real-world interaction. *Psychon Bull Rev*, 1998. 5(4): 644–649.

116 44 percent of people actually notice the gorilla

D. J. Simons and C. F. Chabris, Gorillas in our midst: sustained inattentional blindness for dynamic events. *Perception*, 1999. 28(9): 1059–1074.

117 Conscious of fainter targets . . . also detect targets faster

R. Desimone and J. Duncan, Neural mechanisms of selective visual attention. *Annu Rev Neurosci*, 1995. 18: 193–222.

J. W. Bisley and M. E. Goldberg, Neuronal activity in the lateral intraparietal area and spatial attention. *Science*, 2003. 299(5603): 81–86.

M. Carrasco, C. Penpeci-Talgar, and M. Eckstein, Spatial covert attention increases contrast sensitivity across the CSF: support for signal enhancement. *Vision Res*, 2000. 40(10–12): 1203–1215.

117 Object will be perceived . . . more contrast

M. Carrasco, S. Ling, and S. Read, Attention alters appearance. *Nat Neurosci*, 2004. 7(3): 308–313.

118 Attention emerging from . . . collective neuronal war

Desimone and Duncan, Neural mechanisms.

J. Duncan, EPS Mid-Career Award 2004: brain mechanisms of attention. *Q J Exp Psychol* (Colchester), 2006. 59(1): 2–27.

119 Famous student physics lectures

R. P. Feynman, R. B. Leighton, and M. Sands, *Lectures on physics: complete set*, vols. 1–3. 1998, Boston: Addison Wesley.

122 Ferret brains . . . rewired . . . visual cortex in blind . . . process Braille

J. Sharma, A. Angelucci, and M. Sur, Induction of visual orientation modules in auditory cortex. *Nature*, 2000. 404(6780): 841–847.

L. Von Melchner, S. L. Pallas, and M. Sur, Visual behaviour mediated by retinal projections directed to the auditory pathway. *Nature*, 2000. 404(6780): 871–876.

H. Burton et al., Adaptive changes in early and late blind: a fMRI study of Braille reading. *J Neurophysiol*, 2002. 87(1): 589–607.

124 Constant competition . . . in the brain

Desimone and Duncan (1995), see above.

Duncan (2006), see above.

E. I. Knudsen, Fundamental components of attention. *Annu Rev Neurosci*, 2007. 30: 57–78.

127 Modern version of Phineas Gage

A. R. Damasio, *Descartes' error: emotion, reason, and the human brain.* 1994, New York: HarperCollins.

129 Multiple personality disorder . . . attribute . . . experiences to other personalities

E. Bliss, *Multiple personality, allied disorders, and hypnosis.* 1986, Oxford: Oxford University Press.

131 Aware of your own consciousness

H. Lau and D. Rosenthal, Empirical support for higher-order theories of conscious awareness. *Trends Cogn Sci*, 2011. 15(8): 365–373.

132 Poor at matching their confidence to their accuracy

S. M. Fleming et al., Relating introspective accuracy to individual differences in brain structure. *Science*, 2010. 329(5998): 1541–1543.

135 Three or four conscious items

N. Cowan, The magical number 4 in short-term memory: a reconsideration of mental storage capacity. *Behav Brain Sci*, 2001. 24(1): 87–114; discussion 114–185.

135 "Global workspace theory" proposed by Bernard Baars

B. J. Baars, *In the theater of consciousness: the workspace of the mind.* 1997, New York: Oxford University Press.

B. J. Baars and S. Franklin, An architectural model of conscious and unconscious brain functions: Global Workspace Theory and IDA. *Neural Netw*, 2007. 20(9): 955–961.

136 George Sperling presented subjects

G. Sperling, The information available in brief visual presentations. *Psychological Monographs*, 1960. 74(11, Whole No. 498): 1–29.

137 **Steven Yantis presented subjects with**

S. Yantis, Multielement visual tracking: attention and perceptual organization. *Cogn Psychol*, 1992. 24(3): 295–340.

137 **Our working memory limit . . . same as the monkey's**

E. Heyselaar, K. Johnston, and M. Paré, The capacity limit of the visual working memory of the macaque monkey. *J Vision*. 10(7): 725.

138 **Other species . . . same upper bound . . . honeybee**

H. J. Gross et al., Number-based visual generalisation in the honeybee. PLoS One, 2009. 4(1): e4263.

138 **3 or 4 items . . . maximum that can be practically sustained**

A. Raffone and G. Wolters, A cortical mechanism for binding in visual working memory. *J Cogn Neurosci*, 2001. 13(6): 766–785.

139 **Each holder . . . cope equally well . . . simplest . . . most complex**

J. Duncan, Similarity between concurrent visual discriminations: dimensions and objects. *Percept Psychophys*, 1993. 54(4): 425–430.

140 **Say back a novel sequence that was 80 digits**

K. A. Ericcson, W. G. Chase, and S. Falloon, Acquisition of a memory skill. *Science*, 1980. 208: 1181–1182.

142 **Presented volunteers . . . sequences of 4 double digits**

D. Bor and A. M. Owen, A common prefrontal-parietal network for mnemonic and mathematical recoding strategies within working memory. *Cereb Cortex*, 2007. 17(4): 778–786.

142 **Chess masters . . . remember . . . whole board**

W. G. Chase and H. A. Simon, Perception in chess. *Cogn Psychol*, 1973. 4: 55–81.

143 **Increase the amount of information per item**

G. A. Miller, The magical number seven, plus or minus two: some limits on our capacity for processing information. *Psychol Rev*, 1956. 63(2): 81–97.

144 **Greater the number, the less likely . . . identify each**

C. Bundesen, H. Shibuya, and A. Larsen, Visual selection from multielement displays: a model for partial report, in *Attention and performance XI*, M. I. Posner and M.S.I. Marin, eds. 1985, Hillsdale, NJ: Lawrence Erlbaum, 631–649.

J. J. Todd and R. Marois, Capacity limit of visual short-term memory in human posterior parietal cortex. *Nature*, 2004. 428(6984): 751–754.

144 **Attention has two clear stages**

C. Bundesen, T. Habekost, and S. Kyllingsbaek, A neural theory of visual attention: bridging cognition and neurophysiology. *Psychol Rev*, 2005. 112(2): 291–328.

145 **Leonardo Chelazzi and colleagues used electrodes**

L. Chelazzi et al., Responses of neurons in inferior temporal cortex during memory-guided visual search. *J Neurophysiol*, 1998. 80(6): 2918–2940.

145 **Michael Cohen and colleagues recently carried out**

M. A. Cohen, G. A. Alvarez, and K. Nakayama, Natural-scene perception requires attention. *Psychol Sci*, 2011. 22(9): 1165–1172.

146 **Divert attention away . . . fails to enter consciousness**

J. S. Joseph, M. M. Chun, and K. Nakayama, Attentional requirements in a "preattentive" feature search task. *Nature*, 1997. 387(6635): 805–807.

J. Theeuwes, A. F. Kramer, and P. Atchley, Attentional effects on preattentive vision: spatial precues affect the detection of simple features. *J Exp Psychol Hum Percept Perform*, 1999. 25(2): 341–347.

S. Walker, P. Stafford, and G. Davis, Ultra-rapid categorization requires visual attention: scenes with multiple foreground objects. *J Vision*, 2008. 8(4): 211–212.

146 **Working memory . . . is limited to . . . 4 . . . items**

Cowan (2001), see above.

147 **Psychologists assumed that our working memory capacity was around double this**

Miller (1956), see above.

148 **Douglas Hofstadter's whimsical and influential book**

D. Hofstadter, *Godel, Escher, Bach: an eternal golden braid (20th anniversary edition, with a new preface by the author)*. 1999, London: Penguin.

150 **Prior expectations . . . guide our attention**

L. Melloni et al., Expectations change the signatures and timing of electrophysiological correlates of perceptual awareness. *J Neurosci*, 2011. 31(4): 1386–1396.

C. Summerfield and T. Egner, Expectation (and attention) in visual cognition. *Trends Cogn Sci*, 2009. 13(9): 403–409.

151 **Other species can start to learn . . . grammatical language**

S. Savage-Rumbaugh and R. Lewin, *Kanzi: the ape at the brink of the human mind*. 1994, New York: John Wiley & Sons.

151 **A "language instinct"**

S. Pinker, *The language instinct: the new science of language and mind.* 2003, London: Penguin.

152 **Learning an artificial grammar . . . same brain areas . . . chunking task**

D. Bor et al., Encoding strategies dissociate prefrontal activity from working memory demand. *Neuron*, 2003. 37(2): 361–367.

P. Fletcher et al., Learning-related neuronal responses in prefrontal cortex studied with functional neuroimaging. *Cereb Cortex*, 1999. 9(2): 168–178.

152 **FOXP2**

F. Vargha-Khadem et al., FOXP2 and the neuroanatomy of speech and language. *Nat Rev Neurosci*, 2005. 6(2): 131–138.

154 **The evolutionary advantage of awareness**

D. Bor and A. K. Seth, Consciousness and the prefrontal parietal network: insights from attention, working memory and chunking. *Front Psychol,* 2012. 3: 63.

155 **Stan Beilock and colleagues tested . . . goft-putting**

S. L. Beilock et al., Haste does not always make waste: expertise, direction of attention, and speed versus accuracy in performing sensorimotor skills. *Psychon Bull Rev*, 2004. 11(2): 373–379.

S. L. Beilock et al., When paying attention becomes counterproductive: impact of divided versus skill-focused attention on novice and experienced performance of sensorimotor skills. *J Exp Psychol Appl*, 2002. 8(1): 6–16.

155 **Similar results . . . soccer, baseball . . . touch typing**

B. Castaneda and R. Gray, Effects of focus of attention on baseball batting performance in players of differing skill levels. *J Sport Exerc Psychol*, 2007. 29(1): 60–77.

G. D. Logan and M. J. Crump, The left hand doesn't know what the right hand is doing: the disruptive effects of attention to the hands in skilled typewriting. *Psychol Sci*, 2009. 20(10): 1296–1300.

CHAPTER 5: THE BRAIN'S EXPERIENCE OF A ROSE

160 **Woman was born . . . no cerebellum**

D. Timmann et al., Cerebellar agenesis: clinical, neuropsychological and MR findings. *Neurocase*, 2003. 9(5): 402–413.

160 **Cerebellum . . . 80 percent of . . . brain's neurons**

F. A. Azevedo et al., Equal numbers of neuronal and nonneuronal cells make the human brain an isometrically scaled-up primate brain. *J Comp Neurol*, 2009. 513(5): 532–541.

161 **Isn't enough neural coherence . . . aware . . . when . . . awake**

N. D. Schiff, Recovery of consciousness after brain injury: a mesocircuit hypothesis. *Trends Neurosci*, 2010. 33(1): 1–9.

163 **Similar stories occur for blindtouch**

L. Weiskrantz, *Consciousness lost and found: a neuropsychological exploration*. 1997, Oxford: Oxford University Press.

163 **A visual experience . . . primary visual cortex is missing**

D. H. Ffytche and S. Zeki, The primary visual cortex, and feedback to it, are not necessary for conscious vision. *Brain*, 2011. 134 (Pt 1): 247–257.

163 **Degraded form of consciousness . . . matches their reduced ability**

M. Overgaard et al., Seeing without seeing? Degraded conscious vision in a blindsight patient. PLoS One, 2008. 3(8): e3028.

164 **Blindsight patient in the fMRI scanner**

A. Sahraie et al., Pattern of neuronal activity associated with conscious and unconscious processing of visual signals. *Proc Natl Acad Sci USA*, 1997. 94(17): 9406–9411.

165 **Nikos Logothetis and his team, who carried out**

D. A. Leopold and N. K. Logothetis, Activity changes in early visual cortex reflect monkeys' percepts during binocular rivalry. *Nature*, 1996. 379(6565): 549–553.

N. K. Logothetis, Single units and conscious vision. *Philos Trans R Soc Lond B Biol Sci*, 1998. 353(1377): 1801–1818.

165 **Perception . . . persist . . . momentary gap of the eye blink**

T. J. Gawne and J. M. Martin, Activity of primate V1 cortical neurons during blinks. *J Neurophysiol*, 2000. 84(5): 2691–2694.

G. Rees, G. Kreiman, and C. Koch, Neural correlates of consciousness in humans. *Nat Rev Neurosci*, 2002. 3(4): 261–270.

166 **Primary visual cortex is not quite as dumb**

M. A. Silver, D. Ress, and D. J. Heeger, Neural correlates of sustained spatial attention in human early visual cortex. *J Neurophysiol*, 2007. 97(1): 229–237.

168 **Two other regions also light up at least as brightly**

E. D. Lumer, K. J. Friston, and G. Rees, Neural correlates of perceptual rivalry in the human brain. *Science*, 1998. 280(5371): 1930–1934.

G. Rees, Neural correlates of the contents of visual awareness in humans. *Philos Trans R Soc Lond B Biol Sci*, 2007. 362(1481): 877–886.

168 **Record . . . experiences switch . . . candlestick or faces**

A. Kleinschmidt et al., Human brain activity during spontaneously reversing perception of ambiguous figures. *Proc Biol Sci*, 1998. 265(1413): 2427–2433.

168 **Posterior parietal . . . always activated . . . with the lateral prefrontal**

J. Duncan, EPS Mid-Career Award 2004: brain mechanisms of attention. *Q J Exp Psychol* (Colchester), 2006. 59(1): 2–27.

169 **Stanislas Dehaene . . . showed subjects a rapid sequence**

S. Dehaene et al., Cerebral mechanisms of word masking and unconscious repetition priming. *Nat Neurosci*, 2001. 4(7): 752–758.

169 **Touch or a sound, or . . . combination of senses**

S. Dehaene and J. P. Changeux, Experimental and theoretical approaches to conscious processing. *Neuron*, 2011. 70(2): 200–227.

169 **Physical size . . . brain structure . . . links with awareness**

S. M. Fleming et al., Relating introspective accuracy to individual differences in brain structure. *Science*, 2010. 329(5998): 1541–1543.

R. Kanai, B. Bahrami, and G. Rees, Human parietal cortex structure predicts individual differences in perceptual rivalry. *Current Biol*, 2010. 20(18): 1626–1630.

169 **Antoine Del Cul and colleagues gave patients**

A. Del Cul et al., Causal role of prefrontal cortex in the threshold for access to consciousness. *Brain*, 2009. 132 (Pt 9): 2531–2540.

170 **Jon Simons and colleagues gave a memory test**

J. S. Simons et al., Dissociation between memory accuracy and memory confidence following bilateral parietal lesions. *Cereb Cortex*, 2010. 20(2): 479–485.

170 **Matt Davis and colleagues played volunteers . . . sentences**

M. H. Davis et al., Dissociating speech perception and comprehension at reduced levels of awareness. *Proc Natl Acad Sci USA*, 2007. 104(41): 16032–16037.

170 **The lowest share compared to our primate cousins**

A. A. De Sousa et al., Hominoid visual brain structure volumes and the position of the lunate sulcus. *J Hum Evol*, 2010. 58(4): 281–292.

171 **Wilder Penfield along with his colleague Joseph Evans**

W. Penfield and J. Evans, The frontal lobe in man: a clinical study of maximum removals. *Brain*, 1935. 58: 115–133.

172 **Patients with prefrontal cortex damage . . . working memory deficit**

D. Bor et al., Frontal lobe involvement in spatial span: converging studies of normal and impaired function. *Neuropsychologia*, 2006. 44(2): 229–237.

173 **Hemispatial neglect . . . associated either with . . . prefrontal or parietal**

M. Husain and C. Rorden, Non-spatially lateralized mechanisms in hemispatial neglect. *Nat Rev Neurosci*, 2003. 4(1): 26–36.

173 **Neglected patients . . . place a mark . . . right edge**

A. Parton, P. Malhotra, and M. Husain, Hemispatial neglect. *J Neurol Neurosurg Psychiatry*, 2004. 75(1): 13–21.

173 **Edoardo Bisiach . . . asked hemispatial-neglect patients to imagine**

E. Bisiach and C. Luzzatti, Unilateral neglect of representational space. *Cortex*, 1978. 14(1): 129–133.

174 **Margarita Sarri . . . an experiment on neglect patients**

M. Sarri, F. Blankenburg, and J. Driver, Neural correlates of crossmodal visual-tactile extinction and of tactile awareness revealed by fMRI in a right-hemisphere stroke patient. *Neuropsychologia*, 2006. 44(12): 2398–2410.

175 **"There is no there, there"**

L. C. Robertson, Binding, spatial attention and perceptual awareness. *Nat Rev Neurosci*, 2003. 4(2): 93–102.

175 **Recognize colors and read words, but couldn't do both**

H. B. Coslett and G. Lie, Simultanagnosia: when a rose is not red. *J Cogn Neurosci*, 2008. 20(1): 36–48.

175 **Bob Knight, has come across such a patient**

R. T. Knight and M. Grabowecky, Escape from linear time: prefrontal cortex and conscious experience, in *The new cognitive neurosciences*, M. Gazzaniga, ed. 1995, Cambridge: MIT Press, 1357–1371.

176 **Working memory and attention . . . prefrontal parietal network . . . both**

R. Cabeza and L. Nyberg, Imaging cognition II: an empirical review of 275 PET and fMRI studies. *J Cogn Neurosci*, 2000. 12(1): 1–47.

176 **Increase the number of letters . . . working memory**

T. S. Braver et al., A parametric study of prefrontal cortex involvement in human working memory. *NeuroImage*, 1997. 5(1): 49–62.

176 **Number of abstract relations between items of an IQ task**

J. K. Kroger et al., Recruitment of anterior dorsolateral prefrontal cortex in human reasoning: a parametric study of relational complexity. *Cereb Cortex*, 2002. 12(5): 477–485.

176 **Number of spatial locations you have to remember**

D. Bor, J. Duncan, and A. M. Owen, The role of spatial configuration in tests of working memory explored with functional neuroimaging. *Scand J Psych*, 2001. 42(3): 217–224.

176 **If you switch attention between tasks**

C. Y. Sylvester et al., Switching attention and resolving interference: fMRI measures of executive functions. *Neuropsychologia*, 2003. 41(3): 357–370.

176 **If you attend to visual changes on a screen**

C. Buchel et al., The functional anatomy of attention to visual motion: a functional MRI study. *Brain*, 1998. 121 (Pt 7): 1281–1294.

N. Hon et al., Frontoparietal activity with minimal decision and control. *J Neurosci*, 2006. 26(38): 9805–9809.

176 *Whenever we perform a complex or novel . . . task*

J. Duncan and A. M. Owen, Common regions of the human frontal lobe recruited by diverse cognitive demands. *Trends Neurosci*, 2000. 23(10): 475–483.

176 **These regions . . . closely linked with IQ**

J. R. Gray, C. F. Chabris, and T. S. Braver, Neural mechanisms of general fluid intelligence. *Nat Neurosci*, 2003. 6(3): 316–322.

J. Duncan, A neural basis for general intelligence. *Science*, 2000. 289(5478): 457–460.

177 **Nikki Pratt . . . gave volunteers a classic attentional task**

N. Pratt, A. Willoughby, and D. Swick, Effects of working memory load on visual selective attention: behavioral and electrophysiological evidence. *Front Hum Neurosci*, 2011. 5: 57.

177 **Subsume working memory within an attentional framework**

E. I. Knudsen, Fundamental components of attention. *Annu Rev Neurosci*, 2007. 30: 57–78.

C. Bundesen, T. Habekost, and S. Kyllingsbaek, A neural theory of visual attention: bridging cognition and neurophysiology. *Psychol Rev*, 2005. 112(2): 291–328.

178 **Volunteers would view an array of 16 red squares**

D. Bor et al., Encoding strategies dissociate prefrontal activity from working memory demand. *Neuron*, 2003. 37(2): 361–367.

179 **Same experiment . . . this time with digits**

D. Bor et al., Prefrontal cortical involvement in verbal encoding strategies. *Eur J Neurosci*, 2004. 19(12): 3365–3370.

179 **Moving on to sequences of double digits**

D. Bor and A. M. Owen, A common prefrontal-parietal network for mnemonic and mathematical recoding strategies within working memory. *Cereb Cortex*, 2007. 17(4): 778–786.

181 **Cary Savage and colleagues showed that**

C. R. Savage et al., Prefrontal regions supporting spontaneous and directed application of verbal learning strategies: evidence from PET. *Brain*, 2001. 124 (Pt 1): 219–231.

181 **Vivek Prabhakaran . . . presented letters to participants**

V. Prabhakaran et al., Integration of diverse information in working memory within the frontal lobe. *Nat Neurosci*, 2000. 3(1): 85–90.

181 **Christopher Moore . . . demonstrated that extensive training**

C. D. Moore, M. X. Cohen, and C. Ranganath, Neural mechanisms of expert skills in visual working memory. *J. Neurosci*, 2006. 26(43): 11187–11196.

181 **Stanislas Dehaene . . . showed the transition**

C. Landmann et al., Dynamics of prefrontal and cingulate activity during a reward-based logical deduction task. *Cereb Cortex*, 2007. 17(4): 749–759.

182 **Tammet . . . extreme form of synesthesia**

S. Baron-Cohen et al., Savant memory in a man with colour form-number synaesthesia and Asperger syndrome. *J Consciousness Studies*, 2007. 14(9–10): 237–251.

182 **Tammet . . . far more numbers . . . short-term memory**
Ibid.

182 **Investigate his brain activity . . . one of our chunking tests**

D. Bor, J. Billington, and S. Baron-Cohen, Savant memory for digits in a case of synaesthesia and Asperger syndrome is related to hyperactivity in the lateral prefrontal cortex. *Neurocase*, 2007. 13(5): 311–319.

184 **Convert awkward obstacles into innovative solutions and . . . habits**

D. Bor and A. K. Seth, Consciousness and the prefrontal parietal network: insights from attention, working memory and chunking. *Front Psychol,* 2012. 3: 63.

184 **Crick popularized the idea . . . neurons act in harmony**

F. Crick, *The astonishing hypothesis: the search for the scientific soul.* 1994, New York: Scribner.

184 **Gamma band . . . a frequency previously linked with attention**

H. Tiitinen et al., Selective attention enhances the auditory 40-Hz transient response in humans. *Nature,* 1993. 364(6432): 59–60.

184 **Rats . . . swift . . . synchrony when in a deep sleep**

C. H. Vanderwolf, Are neocortical gamma waves related to consciousness? *Brain Res,* 2000. 855(2): 217–224.

185 **Two . . . labs, Stanislas Dehaene's . . . and Bob Knight's . . . have shown**

R. T. Canolty et al., High gamma power is phase-locked to theta oscillations in human neocortex. *Science,* 2006. 313(5793): 1626–1628.

R. Gaillard et al., Converging intracranial markers of conscious access. PLoS Biol, 2009. 7(3): e61.

186 **Time . . . attention to filter sensory input according . . . goals**

Bundesen et al. (2005), see above.

Gaillard et al. (2009), see above.

187 **Dozens of simultaneous electrodes**

A. Maier, C. J. Aura, and D. A. Leopold, Infragranular sources of sustained local field potential responses in macaque primary visual cortex. *J Neurosci,* 2011. 31(6): 1971–1980.

187 **Victor Lamme's** *recurrent processing model*

V.A.F. Lamme, How neuroscience will change our view on consciousness. *Cogn Neurosci,* 2010. 1(3): 204–220.

188 ***Global neuronal workspace model***

Dehaene and Changeux (2011), see above.

S. Dehaene, M. Kerszberg, and J. P. Changeux, A neuronal model of a global workspace in effortful cognitive tasks. *Proc Natl Acad Sci USA,* 1998. 95(24): 14529–14534.

189 **Studies of how the brain is wired**

D. S. Modha and R. Singh, Network architecture of the long-distance pathways in the macaque brain. *Proc Natl Acad Sci USA, 2010.* 107(30): 13485–13490.

189 **Giulio Tononi's *information integration theory***

G. Tononi, An information integration theory of consciousness. *BMC Neurosci*, 2004. 5: 42.

G. Tononi, Consciousness as integrated information: a provisional manifesto. *Biol Bull*, 2008. 215(3): 216–242.

189 **This theory is similar to two other modern theories**

A. K. Seth et al., Theories and measures of consciousness: an extended framework. *Proc Natl Acad Sci USA*, 2006. 103(28): 10799–10804.

190 **Prefrontal parietal network . . . kind of network . . . high . . . consciousness**

Bor and Seth (2012), see above.

CHAPTER 6: BEING BIRD-BRAINED IS NOT AN INSULT

197 **Story about a mature female bonobo named Matata**

S. Savage-Rumbaugh and R. Lewin, *Kanzi: the ape at the brink of the human mind.* 1994, New York: John Wiley & Sons.

198 **Not . . . conscious . . . between the pressure and the squealing**

D. McFarland, *Guilty robots, happy dogs: the question of alien minds.* 2008, Oxford: Oxford University Press.

198 **Animals appear to get bored**

F. Wemelsfelder, Animal boredom: understanding the tedium of confined lives, in *Mental health and well-being in animals*, F. D. McMillan, ed. 2005, Oxford, UK: Blackwell.

199 **Corvid . . . same brain-to-body ratio as a chimpanzee**

N. J. Emery and N. S. Clayton, The mentality of crows: convergent evolution of intelligence in corvids and apes. *Science*, 2004. 306(5703): 1903–1907.

199 **These birds plan for the future**

C. R. Raby et al., Planning for the future by western scrub-jays. *Nature*, 2007. 445(7130): 919–921.

199 **Scrub jay . . . re-hide the food to fool the observer**

J. M. Dally, N. J. Emery, and N. S. Clayton, Food-caching western scrub-jays keep track of who was watching when. *Science*, 2006. 312(5780): 1662–1665.

200 **Crows can use a sequence of tools**

J. H. Wimpenny et al., Cognitive processes associated with sequential tool use in New Caledonian crows. PLoS One, 2009. 4(8): e6471.

200 **The rooks easily learned . . . drop stones into . . . water**

C. D. Bird and N. J. Emery, Rooks use stones to raise the water level to reach a floating worm. *Current Biol*, 2009. 19(16): 1410–1414.

201 **Observed in the great apes by Daniel Hanus**

D. Hanus et al., Comparing the performances of apes (*Gorilla gorilla, Pan troglodytes, Pongo pygmaeus*) and human children (*Homo sapiens*) in the floating peanut task. PLoS One, 2011. 6(6): e19555.

201 **Five orangutans tested could pass this challenge**

N. Mendes, D. Hanus, and J. Call, Raising the level: orangutans use water as a tool. *Biol Lett*, 2007. 3(5): 453–455.

202 **Other animals that have passed it include**

D. B. Edelman and A. K. Seth, Animal consciousness: a synthetic approach. *Trends Neurosci*, 2009. 32(9): 476–484.

203 **Monkeys can be trained to tell . . . experience flips**

N. K. Logothetis, Single units and conscious vision. *Philos Trans R Soc Lond B Biol Sci*, 1998. 353(1377): 1801–1818.

203 **Nate Kornell . . . showed monkeys a set of dots**

N. Kornell, L. K. Son, and H. S. Terrace, Transfer of metacognitive skills and hint seeking in monkeys. *Psychol Sci*, 2007. 18(1): 64–71.

204 **Posterior parietal cortex . . . activity . . . matches level of confidence**

R. Kiani and M. N. Shadlen, Representation of confidence associated with a decision by neurons in the parietal cortex. *Science*, 2009. 324(5928): 759–764.

204 **Other species that have shown similar skills**

C. Suda-King, Do orangutans (*Pongo pygmaeus*) know when they do not remember? *Anim Cogn*, 2008. 11(1): 21–42.

A. L. Foote and J. D. Crystal, Metacognition in the rat. *Curr Biol*, 2007. 17(6): 551–555.

205 Rats . . . shown to apply simple forms of chunking

T. Macuda and W. A. Roberts, Further evidence for hierarchical chunking in rat spatial memory. *J Exp Psych: Anim Behav Proc*, 1995. 21(1): 20–32.

205 Herbert Terrance trained pigeons to peck

H. S. Terrace, Chunking by a pigeon in a serial learning task. *Nature*, 1987. 325(7000): 149–151.

205 Analogous to human forms of chunking

K. A. Ericcson, W. G. Chase, and S. Falloon, Acquisition of a memory skill. *Science*, 1980. 208: 1181–1182.

205 Extent of our ability to chunk

C. M. Conway and M. H. Christiansen, Sequential learning in non-human primates. *Trends Cogn Sci*, 2001. 5(12): 539–546.

206 Observe different species at play

Ibid.

207 Patricia Greenfield . . . abilities to chunk . . . mirrored . . . language

P. M. Greenfield, Language, tools and brain: the ontogeny and phylogeny of hierarchically organized sequential behavior. *Behav Brain Sci*, 1991. 14: 531–595.

208 Zinacantecos babies and toddlers in southern Mexico

Ibid.

209 Mother's placenta and the fetus . . . safe sedation

C. Koch, When does consciousness arise in human babies? *Sci Am*, September 2, 2009; www.scientificamerican.com/article.cfm?id=when-does-consciousness-arise.

211 Bottlenose dolphins, whose brain weighs . . . 1.8 kilograms

L. Marino, A comparison of encephalization between odontocete cetaceans and anthropoid primates. *Brain Behav Evol*, 1998. 51(4): 230–238.

211 African elephant . . . brain that weighs . . . 6.5 kilograms

M. Goodman et al., Phylogenomic analyses reveal convergent patterns of adaptive evolution in elephant and human ancestries. *Proc Natl Acad Sci USA*, 2009. 106(49): 20824–20829.

211 Sperm whale . . . brain that tops 8 kilograms

L. Marino, Cetacean brain evolution: multiplication generates complexity. *Int J Comp Psych*, 2004. 17(1): 1–16.

211 **Many factors . . . useful ratio of brain to body**

S. Herculano-Houzel, The human brain in numbers: a linearly scaled-up primate brain. *Front Hum Neurosci*, 2009. 3: 31.

212 **Brain . . . similarity . . . in . . . other animals**

Edelman and Seth (2009), see above.

212 **All vertebrates . . . thalamus . . . but not all . . . cortex**

A. B. Butler and W. Hodos, *Comparative vertebrate neuroanatomy: evolution and adaptation*, vol. 2. 2005, Hoboken, NJ: Wiley.

212 **Octopus . . . behaves . . . utterly belie its primitive label**

Edelman and Seth (2009), see above.

213 **If octopuses are conscious . . . never realize it from . . . anatomy**

Ibid.

213 **Giulio Tononi's information integration theory**

G. Tononi, Consciousness as integrated information: a provisional manifesto. *Biol Bull*, 2008. 215(3): 216–242.

214 **Adam Barrett and Anil Seth . . . adapt the theory**

A. B. Barrett and A. K. Seth, Practical measures of integrated information for time-series data. PLoS Comput Biol, 2011. 7(1): e1001052.

215 **Massimini . . . practical rough-and-ready approximation**

M. Massimini et al., Breakdown of cortical effective connectivity during sleep. *Science*, 2005. 309(5744): 2228–2232.

CHAPTER 7: LIVING ON THE FRAGILE EDGE OF AWARENESS

221 **Concussions . . . severe brain damage . . . Alzheimer's disease**

B. Holmes, Deep impact: the bad news about banging your head. *New Scientist*, 2011. 2829: 38–41.

225 **Vegetative state . . . thalamus . . . pathways to the prefrontal cortex**

S. Laureys et al., Restoration of thalamocortical connectivity after recovery from persistent vegetative state. *Lancet*, 2000. 355(9217): 1790–1791.

226 **Rating scales, which standardize and quantify the diagnostic**

C. Schnakers et al., Diagnostic accuracy of the vegetative and minimally conscious state: clinical consensus versus standardized neurobehavioral assessment. *BioMedCentral Neurol*, 2009. 9: 35.

227 **Invalid ... assume ... no conscious life ... completely paralyzed**

M. M. Monti, S. Laureys, and A. M. Owen, The vegetative state. *Brit Med J,* 2010. 341: c3765.

227 **One recent large-scale study ... involved 41 patients**

M. R. Coleman et al., Towards the routine use of brain imaging to aid the clinical diagnosis of disorders of consciousness. *Brain,* 2009. 132 (Pt 9): 2541–2552.

228 **In 2006 with a twenty-three-year-old woman**

A. M. Owen et al., Detecting awareness in the vegetative state. *Science,* 2006. 313(5792): 1402.

228 **Normal, healthy controls performing the same imaginary tasks**

M. Boly et al., When thoughts become action: an fMRI paradigm to study volitional brain activity in non-communicative brain injured patients. *NeuroImage,* 2007. 36(3): 979–992.

229 **Owen ... showed that only about 17 percent**

M. M. Monti et al., Willful modulation of brain activity in disorders of consciousness. *N Engl J Med,* 2010. 362(7): 579–589.

229 **Owen's group ... testing the use of EEG**

D. Cruse et al., Bedside detection of awareness in the vegetative state: a cohort study. *Lancet,* 2011. 378(9808): 2088–2094.

229 **This technique was later used on ... a young Belgian man**

Monti et al (2010), see above.

231 **Davinia Fernandez-Espejo ... relatively novel MRI scanning technique**

D. Fernández-Espejo et al., Diffusion weighted imaging distinguishes the vegetative state from the minimally conscious state. *NeuroImage,* 2011. 54(1): 103–112.

231 **Melanie Boly and colleagues were able to show**

M. Boly et al., Preserved feedforward but impaired top-down processes in the vegetative state. *Science,* 2011. 332(6031): 858–862.

231 **Another principled approach ... uses the TMS-EEG technique**

M. Massimini et al., A perturbational approach for evaluating the brain's capacity for consciousness. *Prog Brain Res,* 2009. 177: 201–214.

232 **Nicholas Schiff ... deep brain stimulation**

N. D. Schiff et al., Behavioural improvements with thalamic stimulation after severe traumatic brain injury. *Nature,* 2007. 448(7153): 600–603.

232 **Raif Clauss . . . a common sleeping drug**

R. Clauss and W. Nel, Drug induced arousal from the permanent vegetative state. *NeuroRehabilitation*, 2006. 21(1): 23–28.

233 **Other research groups . . . shown that zolpidem . . . improve conscious**

J. Whyte and R. Myers, Incidence of clinically significant responses to zolpidem among patients with disorders of consciousness: a preliminary placebo controlled trial. *Am J Phys Med Rehab*, 2009. 88(5): 410–418.

CHAPTER 8: CONSCIOUSNESS SQUEEZED, STRETCHED, AND SHRUNK

237 **(WHO) estimates that up to a quarter of all people**

World Health Organization (WHO), *Global status report on non-communicable diseases 2010*. 2011, Geneva: WHO.

237 **Loss to the economy due to mental illness**

D. E. Bloom et al., *The global economic burden of non-communicable diseases*, W. E. Forum, ed. 2011, Geneva: World Economic Forum.

239 **Autistic children . . . improvements . . . behavioral intervention program**

G. Dawson et al., Randomized, controlled trial of an intervention for toddlers with autism: the early start Denver Model. *Pediatr*, 2009. 125(1): e17–e23.

239 **Autism . . . centrally defined . . . excessive richness of information**

L. Mottron et al., Enhanced perceptual functioning in autism: an update, and eight principles of autistic perception. *J Autism Dev Disord*, 2006. 36(1): 27–43.

H. Markram, T. Rinaldi, and K. Markram, The intense world syndrome—an alternative hypothesis for autism. *Front Neurosci*, 2007. 1(1): 77–96.

239 **Prominent figures . . . suffered from Asperger's syndrome**

H. Muir, Did Einstein and Newton have autism? *New Scientist*, 2003. 2393: 10.

James, I., Singular scientists. *J Royal Soc Med*, 2003. 96(1): 36–39.

240 **Give autistics the Raven's matrices test . . . IQ jumps**

M. Dawson et al., The level and nature of autistic intelligence. *Psychol Sci*, 2007. 18(8): 657–662.

240 **Superior abilities in . . . perceptual and analytical areas**

Mottron et al (2006), see above.

240 **Autism is the opposite of schizophrenia**

B. Crespi, P. Stead, and M. Elliot, Evolution in health and medicine. Sackler colloquium: comparative genomics of autism and schizophrenia. *Proc Natl Acad Sci USA*, 2009.

241 **Small range of superior skills that autistic people adopt**

D. A. Treffert, The savant syndrome: an extraordinary condition. A synopsis: past, present, future. *Philos Trans Royal Soc B: Biol Sci*, 2009. 364(1522): 1351–1357.

241 **". . . [N]umerical comfort blanket"**

D. Tammet, *Born on a blue day: a memoir of Aspergers and an extraordinary mind.* 2006, London: Hodder & Stoughton.

242 **Autistic people have an imbalance of these chemicals**

A. M. Persico and T. Bourgeron, Searching for ways out of the autism maze: genetic, epigenetic and environmental clues. *Trends Neurosci*, 2006. 29(7): 349–358.

242 **Arbaclofen acts to restore this balance**

L. W. Wang, E. Berry-Kravis, and R. J. Hagerman, Fragile X: leading the way for targeted treatments in autism. *Neurotherapeutics*, 2010. 7(3): 264–274.

244 **Sleep deprivation . . . consistently fatal**

C. A. Everson, B. M. Bergmann, and A. Rechtschaffen, Sleep deprivation in the rat: III. Total sleep deprivation. *Sleep*, 1989. 12(1): 13–21.

244 **Sleep . . . a key ingredient in effective learning and memory**

M. P. Walker and R. Stickgold, Sleep-dependent learning and memory consolidation. *Neuron*, 2004. 44(1): 121–133.

244 **Neurons . . . need a period of reduced activity to reset**

G. Tononi and C. Cirelli, Sleep function and synaptic homeostasis. *Sleep Med Rev*, 2006. 10(1): 49–62.

245 **Rise of depression related to . . . severity of the breathing problems**

P. E. Peppard et al., Longitudinal association of sleep-related breathing disorder and depression. *Arch Intern Med*, 2006. 166(16): 1709–1715.

245 **SAD . . . light therapy . . . raise alertness levels**

G. Vandewalle, P. Maquet, and D. J. Dijk, Light as a modulator of cognitive brain function. *Trends Cogn Sci*, 2009. 13(10): 429–438.

246 **David Gozal . . . ADHD if they had sleep apnea**

L. M. O'Brien et al., Sleep and neurobehavioral characteristics of 5- to 7-year-old children with parentally reported symptoms of attention-deficit/hyperactivity disorder. *Pediatr*, 2003. 111(3): 554–563.

246 **REM sleep the next night increases to catch up**

C. F. Reynolds III et al., Sleep deprivation in healthy elderly men and women: effects on mood and on sleep during recovery. *Sleep*, 1986. 9(4): 492–501.

246 **Disturbance in REM sleep . . . depressives and schizophrenics**

D. Riemann, M. Berger, and U. Voderholzer, Sleep and depression—results from psychobiological studies: an overview. *Biol Psychol*, 2001. 57(1–3): 67–103.

M. S. Keshavan, C. F. Reynolds, and D. J. Kupfer, Electroencephalographic sleep in schizophrenia: a critical review. *Compr Psychiatry*, 1990. 31(1): 34–47.

246 **Prefrontal parietal network . . . is . . . underactivated in ADHD**

T. Silk et al., Dysfunction in the fronto-parietal network in attention deficit hyperactivity disorder (ADHD): an fMRI study. *Brain Imaging Behav*, 2008. 2(2): 123–131.

246 **Bipolar disorder . . . ties with sleep**

D. T. Plante and J. W. Winkelman, *Sleep disturbance in bipolar disorder: therapeutic implications. Am J Psychiatry*, 2008. 165(7): 830–843.

247 **Abnormalities in genetic and molecular processes that regulate sleep**

K. Wulff et al., Sleep and circadian rhythm disruption in psychiatric and neurodegenerative disease. *Nat Rev Neurosci*, 2010. 11(8): 589–599.

247 **Coffee intake prevents suicide**

I. Kawachi et al., A prospective study of coffee drinking and suicide in women. *Arch Intern Med*, 1996. 156(5): 521–525.

247 **Increased coffee drinking . . . lower incidence of depression**

M. Lucas et al., Coffee, caffeine, and risk of depression among women. *Arch Intern Med*, 2011. 171(17): 1571–1578.

247 **Light therapy . . . help in all manner of psychiatric conditions**

M. Terman, Evolving applications of light therapy. *Sleep Med Rev*, 2007. 11(6): 497–507.

248 **Poorer quality sleep . . . impair[s] . . . working memory**

M.-R. Steenari et al., Working memory and sleep in 6- to 13-year-old schoolchildren. *J Am Acad Child Adolesc Psychiatry*, 2003. 42(1): 85–92.

248 **Michael Chee . . . lack of sleep . . . less efficient prefrontal parietal**

M.W.L. Chee and W. C. Choo, Functional imaging of working memory after 24 hr of total sleep deprivation. *J Neurosci*, 2004. 24(19): 4560–4567.

248 **Robert Thomas . . . studying adults with sleep apnea**

R. J. Thomas et al., Functional imaging of working memory in obstructive sleep-disordered breathing. *J Appl Physiol*, 2005. 98(6): 2226–2234.

248 **Light therapy . . . boosts . . . attention and working memory**

Vandewalle et al (2009), see above.

248 **ADHD children working memory . . . especially low**

R. Martinussen et al., A meta-analysis of working memory impairments in children with attention-deficit/hyperactivity disorder. *J Am Acad Child Adolesc Psychiatry*, 2005. 44(4): 377–384.

249 **Baba Shiv . . . the difficult . . . number . . . likely to choose . . . cake**

B. Shiv and A. Fedorikhin, Heart and mind in conflict: the interplay of affect and cognition in consumer decision making. *J Cons Res*, 1999. 26(3): 278–292.

249 **Clear relationship . . . poor sleep and obesity**

G. Hasler et al., The association between short sleep duration and obesity in young adults: a 13-year prospective study. *Sleep*, 2004. 27(4): 661–666.

249 **Reduced working memory . . . psychological cause of [schizophrenic symptoms]**

P. Goldman-Rakic, Working memory dysfunction in schizophrenia. *J Neuropsychiatry Clin Neurosci*, 1994. 6(4): 348–357.

H. Silver et al., Working memory deficit as a core neuropsychological dysfunction in schizophrenia. *Am J Psychiatry*, 2003. 160(10): 1809–1816.

J. Lee and S. Park, Working memory impairments in schizophrenia: a meta-analysis. *J Abnorm Psychol*, 2005. 114(4): 599–611.

249 **Schizophrenics also have a dysfunctional prefrontal cortex**

C. S. Carter et al., Functional hypofrontality and working memory dysfunction in schizophrenia. *Am J Psychiatry*, 1998. 155(9): 1285–1287.

250 **Well subjects . . . Ritalin . . . prefrontal functions more efficiently**

M. A. Mehta et al., Methylphenidate enhances working memory by modulating discrete frontal and parietal lobe regions in the human brain. *J Neurosci*, 2000. 20(6): RC65.

250 **Ritalin . . . schizophrenics . . . create psychotic symptoms**

D. S. Janowsky et al., Provocation of schizophrenic symptoms by intravenous administration of methylphenidate. *Arch Gen Psychiatry*, 1973. 28(2): 185–191.

250 **You actually need a medium amount of dopamine**

S. Vijayraghavan et al., Inverted-U dopamine D1 receptor actions on prefrontal neurons engaged in working memory. *Nat Neurosci*, 2007. 10(3): 376–384.

251 **Schizophrenics . . . deficient glutamate . . . knock-on effect on dopamine**

M. Laruelle, L. S. Kegeles, and A. Abi-Dargham, Glutamate, dopamine, and schizophrenia. *Ann NY Acad Sci*, 2003. 1003(1): 138–158.

J. W. Olney and N. B. Farber, Glutamate receptor dysfunction and schizophrenia. *Arch Gen Psychiatry*, 1995. 52(12): 998–1007.

J. Coyle, Glutamate and schizophrenia: beyond the dopamine hypothesis. *Cell Mol Neurobiol*, 2006. 26(4): 363–382.

251 **Normal people, ketamine can . . . turn them into schizophrenics**

P. C. Fletcher and G. D. Honey, Schizophrenia, ketamine and cannabis: evidence of overlapping memory deficits. *Trends Cogn Sci*, 2006. 10(4): 167–174.

251 **Schizophrenics . . . ketamine . . . exacerbate all their symptoms**

A. C. Lahti et al., Effects of ketamine in normal and schizophrenic volunteers. *Neuropsychopharmacol*, 2001. 25(4): 455–467.

252 **About 40 percent of schizophrenics gain any benefit**

S. Leucht et al., How effective are second-generation antipsychotic drugs? A meta-analysis of placebo-controlled trials. *Mol Psychiatry*, 2008. 14(4): 429–447.

252 **67 percent of schizophrenic patients report [adverse side effects]**

J. A. Lieberman et al., Effectiveness of antipsychotic drugs in patients with chronic schizophrenia. *New Engl J Med*, 2005. 353(12): 1209–1223.

252 **Anecdotal evidence . . . schizophrenics function better . . . never been prescribed**

K. Hopper and J. Wanderling, Revisiting the developed versus developing country distinction in course and outcome in schizophrenia: results from ISoS, the WHO collaborative followup project. *Schizophr Bull*, 2000. 26(4): 835–846.

252 **Sandeep Patil . . . created a drug that targets**

S. T. Patil et al., Activation of mGlu2/3 receptors as a new approach to treat schizophrenia: a randomized Phase 2 clinical trial. *Nat Med*, 2007. 13(9): 1102–1107.

253 **Single genes . . . turned on or off . . . sections of DNA . . . change neurotransmitter function**

F. Holsboer, How can we realize the promise of personalized antidepressant medicines? *Nat Rev Neurosci*, 2008. 9(8): 638–646.

254 **Genes coding for dopamine and prefrontal function . . . psychiatric conditions**

A. Meyer-Lindenberg and D. R. Weinberger, Intermediate phenotypes and genetic mechanisms of psychiatric disorders. *Nat Rev Neurosci*, 2006. 7(10): 818–827.

254 **Genes . . . respond to SSRIs . . . pass the blood-brain barrier**

Holsboer (2008), see above.

254 **Novel brain-scanning . . . measure neurotransmitter changes**

P. M. Matthews, G. D. Honey, and E. T. Bullmore, Applications of fMRI in translational medicine and clinical practice. *Nat Rev Neurosci*, 2006. 7(9): 732–744.

255 **Healthy nonelderly people don't gain any generalized improvement**

A. M. Owen et al., Putting brain training to the test. *Nature*, 2010. 465(7299): 775–778.

255 **Cognitive training is useful in staving off dementia**

R. S. Wilson et al., Cognitive activity and the cognitive morbidity of Alzheimer disease. *Neurol*, 2010. 75(11): 990–996.

255 **Torkel Klingberg . . . gave ADHD children . . . tasks to practice**

T. Klingberg, H. Forssberg, and H. Westerberg, Training of working memory in children with ADHD. *J Clin Exp Neuropsychol*, 2002. 24(6): 781–791.

255 **Training in the scanner . . . boost prefrontal parietal . . . activity**

P. J. Olesen, H. Westerberg, and T. Klingberg, Increased prefrontal and parietal activity after training of working memory. *Nat Neurosci*, 2004. 7(1): 75–79.

255 **Joni Holmes . . . compared . . . Ritalin . . . to working memory training**

J. Holmes et al., Working memory deficits can be overcome: impacts of training and medication on working memory in children with ADHD. *Appl Cogn Psychol*, 2010. 24(6): 827–836.

255 **Melissa Fisher . . . cognitive-training program . . . schizophrenic patients improved**

M. Fisher et al., Neuroplasticity-based cognitive training in schizophrenia: an interim report on the effects 6 months later. *Schizophr Bull*, 2010. 36(4): 869–879.

256 **Sylvia Nasar's book**

S. Nasar, *A beautiful mind: a biography of John Forbes Nash, Jr.* 1998, New York: Simon & Schuster.

257 **Half-strangled prefrontal cortex can . . . suppress amygdala activity**

S. J. Bishop, Neurocognitive mechanisms of anxiety: an integrative account. *Trends Cogn Sci*, 2007. 11(7): 307–316.

A.F.T. Arnsten, Stress signalling pathways that impair prefrontal cortex structure and function. *Nat Rev Neurosci*, 2009. 10(6): 410–422.

260 **Depressive or anxious patients . . . fearful stimuli in the scanner**

J. S. Greg et al., Increased amygdala and decreased dorsolateral prefrontal BOLD responses in unipolar depression: related and independent features. *Biol Psychiatry*, 2007. 61(2): 198–209.

S. J. Bishop, Trait anxiety and impoverished prefrontal control of attention. *Nat Neurosci*, 2009. 12(1): 92–98.

260 **Period of stress . . . turn off prefrontal function**

Arnsten, Stress signalling pathways.

262 **Meditative state increases activity in the prefrontal parietal network**

B. R. Cahn and J. Polich, Meditation states and traits: EEG, ERP, and neuroimaging studies. *Psychol Bull*, 2006. 132(2): 180–211.

262 **Regular meditation . . . permanently change the prefrontal parietal network**

J. A. Brefczynski-Lewis et al., Neural correlates of attentional expertise in long-term meditation practitioners. *Proc Natl Acad Sci USA*, 2007. 104(27): 11483–11488.

262 **Long-term meditation shifts the see-saw battles**
Ibid.

262 **Long-term meditation increases the thickness of the prefrontal cortex**

S. W. Lazar et al., Meditation experience is associated with increased cortical thickness. *Neuroreport*, 2005. 16(17): 1893–1897.

262 **Two months of meditation . . . shrink the . . . amygdala**

B. K. Holzel et al., Stress reduction correlates with structural changes in the amygdala. *Soc Cogn Affect Neurosci*, 2010. 5(1): 11–17.

262 Long-term meditation does improve a range of . . . tasks

M. Kozhevnikov et al., The enhancement of visuospatial processing efficiency through Buddhist deity meditation. *Psychol Sci*, 2009. 20(5): 645–653.

A. Moore and P. Malinowski, Meditation, mindfulness and cognitive flexibility. *Conscious Cogn*, 2009. 18(1): 176–186.

H. A. Slagter et al., Mental training affects distribution of limited brain resources. PLoS Biol, 2007. 5(6): e138.

262 Meditation . . . reduce a person's need for sleep

P. Kaul et al., Meditation acutely improves psychomotor vigilance, and may decrease sleep need. *Behav Brain Funct*, 2010. 6: 47.

263 Four meditation sessions . . . reduce . . . tiredness . . . increase working memory

F. Zeidan et al., Mindfulness meditation improves cognition: evidence of brief mental training. *Conscious Cogn*, 2010. 19(2): 597–605.

263 Yi-Yuan Tang . . . five days were needed . . . improve on an attentional task

Y. Y. Tang et al., Short-term meditation training improves attention and self-regulation. *Proc Natl Acad Sci USA*, 2007. 104(43): 17152–17156.

263 Meditation . . . effective weapon against . . . a host of . . . conditions

J. N. Teasdale et al., Prevention of relapse/recurrence in major depression by mindfulness-based cognitive therapy. *J Consulting Clin Psychol*, 2000. 68(4):615–623.

J. Kabat-Zinn et al., Effectiveness of a meditation-based stress reduction program in the treatment of anxiety disorders. *Am J Psychiatry*, 1992. 149(7): 936–943.

J. Kabat-Zinn, L. Lipworth, and R. Burney, The clinical use of mindfulness meditation for the self-regulation of chronic pain. *J Behav Med*, 1985. 8(2): 163–190.

D. P. Johnson et al., Loving-kindness meditation to enhance recovery from negative symptoms of schizophrenia. *J Clin Psychol*, 2009. 65(5): 499–509.

Z. V. Segal et al., Antidepressant monotherapy vs sequential pharmacotherapy and mindfulness-based cognitive therapy, or placebo, for relapse prophylaxis in recurrent depression. *Arch Gen Psychiatry*, 2010. 67(12): 1256–1264.

Illustration Credits

Illustrations follow page 158.

Figure 1   Phineas Gage's skull and brain: Wikimedia Commons. Image originally appeared in John Martyn Harlow, "Recovery from the passage
of an iron bar through the head," *Publications of the Massachusetts
Medical Society* 2 (1868): 327–347.

Figure 2   Examples of complexity: Dow Jones, Wikimedia Commons, permission granted under GNU Free Documentation License by K.
Boroshko, http://commons.wikimedia.org/wiki/File:Finance-dow
jones-chart1.jpg. Ants, Jacinda Brown, used with permission. Galaxy,
courtesy of National Aeronautics and Space Administration. Fractal,
Wikimedia Commons.

Figure 3   Schematic of the human brain (base image): istockphoto.com.

Figure 4   The four lobes of the human brain: Centre for Neuro Skills,
Neuroskills.com, used with permission.

Figure 5   An example of change blindness: Copyright Ronald Rensink, used
with permission.

Figure 6   Examples of stimuli that induce repeated switches in visual perception: Reprinted from *Neuron*, vol. 21(4), F. Tong, K. Nakayama,
J. T. Vaughan, and N. Kanwisher, Binocular rivalry and visual awareness in human extrastriate cortex, 753–759, copyright (1998), with
permission from Elsevier.

Figure 7   Three types of cup assembly: Reprinted from *Trends in Cognitive
Sciences*, vol. 5(12), Christopher M. Conway and Morten H. Christiansen, Sequential learning in non-human primates, 539–546, copyright (2001), with permission from Elsevier.

Figure 8   In the fMRI scanner: Reprinted from *Neuron*, vol. 37(2), Daniel Bor,
John Duncan, Richard J. Wiseman, and Adrian M. Owen, Encoding
strategies dissociate prefrontal activity from working memory demand, 361–367, copyright (2003), with permission from Elsevier.

Figure 9   CT scan comparison of a normal brain and Terri Schiavo's brain:
Originally released to the public domain by Terri Schiavo's doctor,
Dr. Robert Cranford.

Index

Abnormalities
  brain anatomy, 253–254
  genetic, 240, 246–247, 249
  sleep, 243–247
Absentmindedness, 110, 111, 115
Aczel, Balazs, 94
Adaptability, complexity and, 42, 45, 55
Algorithms, 30, 76, 215
Alzheimer's disease, 221. *See also* Dementia
Amino acids, 43, 46, 47
Amygdala, 85–86, 257, 258, 260, 262
  stress and, 257, 258, 260
  prefrontal cortex and, 257, 262
Anesthesia, 80–81, 87–88, 90–91
  learning and, 88–91
  unconsciousness and, 88, 91
Animal
  chunking, 205–208
  consciousness, 198–208, 210–216
  ethics, 216–218
  intelligence, 199–201
  self-knowledge, 203–204
  self-recognition, 202–203
  tool use, 200–201
Antipsychotics, 250–251, 252
Anxiety, 237, 247, 260, 262, 263
Apes, 201, 216, 217
Arbaclofen, 242, 252

Artificial intelligence, 76, 210, 216, 218
Asperger's Syndrome, 182, 239, 241. *See also* Autism
Attention, 117, 134, 146, 156, 176, 267
  boosting, 84, 112, 116–118, 126–127, 134–135, 138–139, 144
  brain and, 118, 177
  consciousness and, 125–127
  filtering, 112–113, 116, 118, 123–124, 126–127, 134–135, 161, 177, 186, 260, 263
  focus, 144, 264
  information processing and, 161
  working memory and, 136, 176, 262
Attention deficit hyperactivity disorder (ADHD), 245, 247, 249, 254
  cognitive training and, 255
  prefrontal parietal network and, 246
  Ritalin treatment for, 250–251
  working memory and, 248, 255
Autism, 182, 238–242, 250
  behavior therapy, 239, 240, 242
  neurotransmitters and, 242
  opposite of schizophrenia, 240
  over-consciousness and, 238–242
Awareness. *See* Consciousness

Baars, Bernard, 135, 136, 188
Bacteria, xii, 45, 46, 49–51, 59, 65, 67,
 72–73, 82
 absence of consciousness and, 48
 antibiotic resistance in, 55
 evolution and, 63–64
 information processing by, 62, 81,
  88
 innovations in, 57
 natural selection and, 64
Bálint's syndrome, 175
Baron-Cohen, Simon, 239
Barrett, Adam, 214–215
Bauby, Jean-Dominique, 83
Behavior, 16, 155, 270
 animal, 70, 74, 196, 197, 198, 205,
  210
 assessment of in vegetative state,
  226, 227, 231
 chemical signaling and, 10
 consciousness and, 169, 198, 211,
  218–219
 controlling, 255
 emergent, 10
 enhancing, 71, 154
 flaws in, 173, 248, 268–270
 information and, 68, 206
 modifying, 98
 movement and, 68–69
 repetitive, 238, 239
 stress and, 198
 subliminal messages and, 99
 treatment, 239, 254
Beilock, Sian, 155
Beliefs
 brain-based, 65, 70, 74
 evolutionary equivalent of, 30,
  39–41, 44, 48, 54, 59, 63, 153
 irrational or superstitious, 74,
  97–98, 100, 129, 250
Binocular rivalry, 165, 168, 203

Biological machines, 11, 47, 51, 63, 70,
 73
Bipolar disorder, 235, 236, 246, 247
Bisiach, Edoardo, 173
Blade Runner (film), 28
Blindsight, 162–164, 165, 168, 173
Blindtouch, 163
Boly, Melanie, 231
Bonobos, 195–198, 206, 207, 208, 212,
 216
Born on a Blue Day (Tammet), 241
Braille-reading, visual cortex
  activated, 68, 122
Brain, 151
 anatomy of, 22, 213
 artificial, 28, 30
 attention and, 118
 chemistry, 1, 253
 complex, 10, 24, 65, 220, 244, 254,
  265
 computational landscape of, 66–72,
  76
 computers and, 2, 11, 17, 19, 22, 25,
  76
 consciousness and, x-xii, xvi, 1, 6,
  10, 13, 25, 33, 157, 158–159, 160,
  162, 169, 173–174, 184, 191, 198,
  210–212, 214–215, 227
 development of, 75, 240, 261
 evolution and, 23, 68–69, 81–86,
  265, 266
 function of, 158, 211
 information processing and, 119,
  244
 interconnections of, 85, 104, 176,
  188–189
 language and, 158
 mapping, 26n, 118
 mind and, 7, 9
 parallel/probabilistic nature of, 22,
  24, 25

processing by, 7, 66n, 82, 118, 134

recovery of, 222, 223, 231

reptilian, 82, 85, 86, 160

rhythms/waves, xvi, 81, 86–87, 138n, 177, 184–187

sensory regions of, 163

silicon, 11, 27

size of, 211, 212

specialized regions of, 82–85, 158, 190

structure of, 82–84, 211

training, 255–256

tumors, 83, 171, 221

visual, 166

weight of, 113n, 211, 211n

Brain activity, 4, 6, 25, 29, 86, 164, 227, 228, 248

chunking and, 182–183

communication by, 229–230

Brain damage, 8, 128, 172, 174, 233, 234, 267

consciousness and, xvii, 158, 220–225

risk of, 222

Brain network, 176, 188–189

Brain-scanning, 1–3, 6, 7, 168, 176–178, 227, 229, 254. *See also* Brain activity; fMRI

Brain stem, 82–83, 86, 160

Breathing, xiv, 83, 86, 245

Bulimia, 223, 224

Bush, George W., 224

Bush, Jeb, 224

Byron, Lord, 109

Caenorhabditis elegans, 50–51, 70, 71, 214

Cancer, 237

Cardiovascular disease, 237

Cerebellum, 160, 190

Chabris, Christopher, 115

Change blindness, 114–116

Changeux, Jean-Pierre, 188

Chaos, xii, 40–42

misinterpreting, 104

stability and, 41, 44–45, 47, 154

Chee, Michael, 248

Chelazzi, Leonardo, 145

Chemical

brain, 120–122. *See also* Neurotransmitters

complexity, breeding, 37–40

constituents of life, 38, 39, 42, 44

imbalances, consciousness and, 250–254

signaling, 10, 73, 121

Chimpanzees, 195–198, 201, 202, 205, 206, 207, 208, 216

brain of, 212

Chinese Room argument, 17–21, 23, 25–28

Choices. *See* Decision Making/decisions

Choo, Wei Chieh, 248

Chromosomes, 49, 57

Chunking, 152n, 178–184, 207n, 241. *See also* Chunks; Strategies; Structure

animal, 205–208

arts, use of, 148, 149

brain activity and, 182–183

consciousness and, 146–151, 152, 270

destructive form of, 259

fruits of, 152–156

hierarchical, 152, 208

language and, 151–152

learning improvement and, 138–143

levels of, 205–206

mathematical, 180

memory-based, 179, 180, 181

Chunking (*continued*)
  prefrontal parietal network and,
    179, 181
  process of, 146–147, 150, 154, 183,
    193
  propensity for, 258–259
  strategies for, 136, 138, 140–141,
    143, 147, 179–181, 205
  visual, 167
Chunks, 144, 183
  building, 149, 156, 268, 269
  deep-seated, 259
  detecting, 149, 156
  information, 147, 151, 269
  innovative, 154
  language, 152
  mathematical, 180
  memory, 143, 150, 179
  mental, 186, 262, 269
  questioning, 270
  spotting, 182
  unconscious mind and, 153
Claus, Ralf, 232
Clayton, Nicky, 199, 200
Coffee: protection against depression,
    247
Cognition, 98, 125, 154, 157, 171, 183,
    213, 263. *See also* Thought
  consciousness and, 208
Cognitive training, 255, 255n, 264
Cohen, Michael, 145, 181
Coma, 160, 226, 233
Communication, xi, 151, 162, 240
  brain activity and, 229–230
  information, 62
  language, 28
  neural, 88, 184, 253
  two-way brain, 188
Complexity, 31, 63, 71–72, 75, 107,
    138–143, 221–223
  adaptability and, 42, 45, 55

  flexibility and, 44
  neural, 253
Computation, 76, 191
  biological, 23–24, 61
Computers, 21–22, 27, 30, 43n, 59, 67,
    76, 97, 162, 259
  biological, 4, 75, 266
  brain and, 11, 17, 19, 22, 25, 76
  Chinese Room argument and,
    17–21, 23, 25–28
  consciousness and, 15–21, 23,
    25–28
  emotions/sensations and, 15–16
  evil, 5
  impact of, 61
  language and, 17–20
  meaning, ability to process, 17–21,
    23, 25–28
  probabilistic parallel, 23
  serial deterministic, 23
  silicon, 26, 28, 159
  thoughts and, 29
Concept of Mind, The (Ryle), 8
Concepts, xii, 9, 128
  layering of, 134
  hierarchical, 206
Concussions, 221, 222
Conjoined twins, 28–29, 129, 191
Consciousness
  anesthesia and, 79–81, 87–88, 90–
    91
  animal, 199, 202–203, 216, 218
  artificial, 25–28, 218
  attention and, 116–118, 125–127.
    See also Attention
  attentional enhancement of,
    117–118
  biological form of, 17
  blindsight and, 162–164, 165, 168,
    173

brain damage and, xvii, 8, 128, 172, 174, 233, 234, 267
brain damage, detecting in, 225–231
brain damage, repairing, 232–234
brain regions for, 149, 159–161, 162–184, 189. *See also* Prefrontal parietal network; Thalamus
brain waves/rhythms for, 184–186
building, xviii, 134, 191, 209, 254–256
capacity for, xvi-xvii, 108, 111, 190, 212
changes in, 6, 7, 158, 236
characteristics of, 15
chunking and, 146–151, 152, 270
complexity of, 138–143
components of, 122, 128, 134–138, 158, 214
computational nature of, 11, 15–16, 30. *See also* Chinese Room argument
contents of, 145
continuum, 211, 219
decision making and, 98–108
driving force of, 127
dysfunctional, 220
early, 209
emotions/sensations and, 15–16, 127–128
engaging, 154, 183
essence of, 12, 189, 265
evolutionary advantage of, 154
evolutionary expansion in humans, 208, 219
explaining, 14, 158, 187, 193–194, 229
extensive, xviii, 81, 129, 191, 204, 217
fetus and questions of, 209
fragility of, 266

forms of, 127, 203, 218
healing, 263–264
higher-order, 131, 133, 134
human compared with other species, 202, 204
infant, 209
layers of, 128–135
level of, xvi, 29, 81, 91, 128, 131, 202–203, 213–216, 217, 218, 219, 227, 228, 267, 269
limits of, 114–116, 128, 143–146
location of. *See* Consciousness: brain regions for
loss of, 80, 83, 86, 128, 174, 226
meaning and, 19
measuring, 210–212, 215, 216
meditation and, 260–263
mental illness and, 235–264
minimal, 81, 98, 226–229, 231–233
models. *See* Consciousness: theories
mystery of, 7, 30, 32, 158, 184
neuronal signature of, 185
neurotransmitters and, 250–254
overabundance of, 238, 239
prefrontal parietal network and, 169, 171, 174, 176–184, 185–191, 194, 204, 212, 231, 246, 248
as physical event, x, 15. *See also* Mind-body duality
purpose of, 146, 147, 152, 158, 161, 177, 185–186, 188, 193, 238, 266
qualitative vs. quantitative, 218–220
quantifying, 189–193, 213–216, 231
reducing, 251
scanning, 168–171
science of, xi, xvii, 4, 31, 32, 33, 184, 216, 230, 265, 268, 269, 270

Consciousness (*continued*)
self-, 98, 127–134, 129, 172, 202, 219, 259. *See also* Mirror-recognition test
signature of, xvi, 187
stress and, 256–260
subjectivity of, 7, 11, 28–31, 191
thalamus and, 29, 84, 160–162, 166, 186, 189, 194, 212–213, 225, 231, 232
theories of, xvi, 15, 131, 135–136, 143, 186–193, 215
timescale of, 104, 186
unconscious mind and, 82, 92, 97
utility of, 133
visual cortex and, 164–168
working memory and, 135–147, 150–151, 153–155, 159, 175–181, 183, 185–188, 193, 208, 219, 246–252, 254–256, 260, 262–264
Conscious control, 257, 266
boosting, 255, 256
limited, 267, 268
Cooperation in biology, 56–61
Cortex, 69, 84, 125, 126, 127, 145, 161, 167, 168, 174, 175, 185, 186, 187, 190, 215, 231
consciousness and, x, 153
general-purpose, 178
modern, 86
motor, 2
neurons and, 126
rebuilding, 233–234
stress and, 257
Corvids, 199, 202, 217
Crick, Francis, 184
Cronin, Alison, 195
Cronin, Jim, 195
Crows, 198–201. *See also* Corvids
tools and, 200–201, 201n

Culture, 17, 18, 92n, 99
Curiosity, xv, 71, 72, 155–156

Damasio, Antonio, 127
Dante, 52
Darwin, Charles, 57
Data, 50, 151, 158, 187
analysis, xii, 112–114, 116–118, 126–128, 134–135, 138–139, 144, 146, 155, 163. *See also* Information processing
compression of, 140, 149–150. *See also* Chunking
filtering, 111–112, 134. *See also* Attention
patterns/structure and, 149
sensory, 163
Davis, Matt, 170
Dawkins, Richard, 54, 57, 58, 216
Daydreams, 80, 109–111
Death, 49–53, 54–56, 83
barely dissociable from life, 225–226
Decision making/decisions, 92, 93, 98–108, 112, 125, 131, 267, 270
CIA and, 99, 99n
conscious, 99, 100, 105–108
freedom of, 102–106. *See also* Free will
knowing accuracy of perceptual, 203, 204
thoughts and, 102
unconscious, 99, 100, 101, 102–106
Deep brain stimulation, 232
Dehaene, Stanislas, 169, 181, 185, 188, 189
Del Cul, Anthony, 169–170
Delusions, xviii, 250, 252
Dementia, 247, 255n
Dennett, Daniel, 103–104

Deoxyribonucleic acid (DNA), xvi, 37, 41, 49, 50–51, 54, 57, 60, 62, 63, 65, 82, 153, 240, 265, 267. *See also* Genes
  bacteria and, 64
  ideas in, xii, 194
  inflexibility of, 47
  junk, 57, 253
  language of, 45–47, 122
  protein and, 45, 88
  random mixing of, xiii
  RNA and, 43, 43n, 46
  sections of, 61
  viral donation/modification of, 55–56
Depression, 236, 237, 245, 247, 260, 262, 263
  cognition and, 263
  sleep and, 245, 247
Descartes, René, xi, 10, 11
  consciousness and, 28
  Ghost in the Machine and, 9
  Cogito ergo sum of, 4, 5
  mind-body duality and, 4–8
  subjectivity and, 7, 11, 28, 29
Descartes' Error (Damasio), 127
Desimone, Bob, 114
Development, 75
  of chunking skills, 206–208
  delays in autism, 238–242
Diffusion tensor imaging, 231
Digit sequences
  chunking, 142
  limits in working memory, 144
  memorizing long, 140–141, 141n, 147, 150
  rehearsal impairing food decisions, 249
  subjects recalling during fMRI, 179
  synesthesia and, 182

Tammet, Daniel, recalling during fMRI, 182–183
Dijksterhuis, Ap, 91–92, 94, 94n, 96
Discourse on the Method (Descartes), 5
Divine Comedy (Dante), 52
Diving Bell and the Butterfly, The (Bauby), 83
Dolphins, 211, 212, 217
Dopamine, 250, 251, 252, 254
Drives, conscious or unconscious, 113, 258, 268, 269–270. *See also* Goals; Motivations
Drugs, 250–254, 264. *See also* Anesthesia; Antipsychotics; Arbaclofen; Coffee; Ritalin; Propofol; SSRIs; Zolpidem
  addictive, 245
  anesthetic, 80
  antipsychotic, 250–251
  combining, 253
  ineffective, 235
  overdose, 225
Dualism. *See* Mind-body duality

Early Start Denver Model, 239
Echolocation, 12, 12n
Edelman, David, 213
Edelman, Gerald, 189n
Einstein, Albert, xv, 13, 36, 239
Electroencephalography (EEG), 87, 88, 103, 184, 185, 186, 215, 229, 231
Elephants, 202, 217, 258
  brain weight of, 211
Elle magazine, Bauby and, 83
Emergentism, 10, 126, 127n
Emotions, 17, 73, 131, 202, 220, 238, 257, 263, 270
  complex, 31, 71–72

Emotions (*continued*)
  computers and, 15–16
  depicting, 16
  evolutionary underpinnings of, 71, 72
  handling, 262
  orbitofrontal cortex and, 85
  overestimating value of, 127–128
  signals from, 71, 113
  unawareness of, 132
Energy (metabolic demands), 105, 113, 113n
  conscious long-term saving of, 153–154
Energy levels, feeling of, 236, 245
Environment, 65, 71, 72, 238
  changing, 38, 48
  control over, 73, 75
Epilepsy, 185
Ericsson, K. Anders, 140
Ether, 35–36
Ethical questions, consciousness and, 216–218
Evans, Joseph, 171
Evolution, xii, 38, 42, 50, 51, 57, 58, 59, 60–61, 70, 71, 72, 75–77, 138n, 121, 128, 198, 218, 267
  animals and, 66
  bacteria and, 63–64
  brain and, 23, 68–69, 81–86, 265, 266
  consciousness and, 193, 265
  environmental niches and, 82
  essence of, 37
  fear and, 84
  genetic hypothesis-testing via, 65
  hardwiring and, 89
  information-processing and, 73
  internal, 63–66, 66n
  learning and, 63
  lesson of, 170–171

  replication and, 53
  scientific method and, 36
Evolutionary pressure, 55, 56, 57, 59, 61
Experience. *See* Consciousness

Faces, perception of, 25, 26, 59, 84, 159, 165, 168, 185, 187
  fusiform face area (FFA) and, 25, 185, 187
Fears, 84, 86. *See also* Anxiety
Federer, Roger, 75, 76
Fedorikhin, Alexander, 249
Feelings. *See* Emotions
Fernandez-Espejo, Davinia, 231
Feynman, Richard, 118–119
Feynman Lectures on Physics, The (Feynman), 119
Fiorito, Graziano, 213
Fisher, Melissa, 255
Fleming, Steve, 132
Flexibility, 120–121, 219
  complexity and, 44
  stability and, 42
fMRI studies, 2, 61, 105, 158, 164, 168, 169, 170, 174, 178, 179, 184, 221, 227, 229, 248
Fovea, 117
Free will, 101, 103–104, 106–108, 107n
Frontal lobes, 8, 85, 171. *See also* Prefrontal cortex
Fusiform face area (FFA), 25, 185, 187

Gage, Phineas, 7–8, 85, 127, 128
Gambling based on self-knowledge, 203–205
Gamma-amino butyric acid (GABA), 86, 242
Gamma rhythms, consciousness and, 184–187
  high gamma variant, 185–186

Gazzaniga, Michael, 157–158
Genes/Genetics, 37, 43, 48, 55–56, 65, 69, 75, 107, 153
  coding, 54, 62n, 254
  collections of, 57, 61
  combinations of, 59
  control, 73
  disorders of, 58, 220, 245, 246, 249, 251
  Hox, 62
  as ideas, 58
  interaction of, 59, 60
  language of, 45–47
  learning and, 44
  mutation-causing, 58
  similarities between consciousness and, 154
Genetic disorders, 58, 220
Genius cells, 61–63
Genomes, 45, 56, 57, 59, 62n
Ghost in the machine, 7–10
Girl with the Dragon Tattoo, The (Larsson), 210
Gladwell, Malcolm, 92n
Global neuronal workspace model, 188–190
Global workspace theory, 135–136, 188
Glutamate, 251, 252, 253
Goals, 102, 143, 266
  biological, 126
  conscious minds and, 102
  consciousness and, 153
  unconscious, 107
Gödel, Escher, Bach: An Eternal Golden Braid (Hofstadter), 148n
Goodall, Jane, 216
Gorillas, 202, 216
  brain of, 212
Gozal, David, 246

Grammar
  artificial, 95–96, 152
  natural, 90, 95, 151, 207n
Great Ape Project, 216
Greenfield, Patricia, 207n

Habits, 74, 181, 184, 271
Hallucinations, 250, 252
Hanus, Daniel, 201
Harlow, John, 8
Haynes, John-Dylan, 105
Heart disease, 237
Heart rate, 71, 83
Hebb's law, 121
Hemispatial neglect. *See* Neglect
Hobbes, Thomas, 7
Hofstadter, Douglas, 148n
Hogan, Krista and Tatiana (conjoined twins), 28–29, 129, 191
Holmes, Joni, 255

Ideas, 11, 40, 42, 66, 194. *See also* Beliefs
  biological, 64
  DNA-based, xvi, 59
  levels of in nature, 56–61
  mental pyramid of, 134
  storing, 194
Identity
  confusable, 129
  genetic, 50, 101
  lost due to stroke, ix, x
Imagination, 4, 11, 53, 119, 173–174, 201, 229
Inferotemporal cortex, 144–145, 167, 168, 189
Information, 93, 116, 119, 136, 139, 161, 162–163, 168, 178, 187, 190, 271
  acquiring, 15, 67, 95, 97, 193–194
  analysis of, 61

Information (*continued*)
atoms of, 56
brain process, 59
carrying capacity, 44
color based, 15, 16
combined, 192
competing, 124
complex, 95
compressing, 149
consciousness and, 36–37, 39, 81, 89, 95, 142, 163, 188, 194, 241
distribution of, 121
fidelity of, 40
flow of, 120, 123
hierarchy, 60
increase in, 142
input of, 112
letters of in genetics, 44
lower-level, 214
manipulating, 91
neurons and, xiii, 24–25, 124, 194
physical structure of primordial life and, 39–40
quantity of, 42, 194
redundancy in, 149. *See also* Chunking
semi-chaotic activity and, 40–41
sequence of, 44
sharing, 29–30, 122, 187
sources of, 118, 124
storing, 39, 43n, 44, 46
streaming, 53, 77, 114, 142
visual, 117, 124, 165
Information integration theory, 189–193, 189n, 213, 214, 219
Information processing, xiii, xv, 23, 25, 31, 40, 46, 47, 50, 62, 65, 66, 68, 73, 75, 84, 106, 111, 112, 118, 134, 154, 189, 256
attention and, 161
brain and, 86, 119, 244

consciousness and, xii, 81, 88, 90, 127
enhancing, 117
levels of, 56–61
unconsciousness and, xiv
Information structure, xv, 56, 57, 90, 148, 152, 208. *See also* Structure
Innovations, xvi, 51, 54, 57, 58, 74, 147, 154, 184, 236, 259–260, 269
chaotic, 153
consciousness and, xiii, xv, xvii, 48, 102, 216–217, 220, 258
desperate times and, 47–49
evolutionary, 56
maximal, 48
primate, 49
spurious, xviii
tools of, 266
Insanity, 256–257
Instinct, 84, 113
language, 151, 152n
Intellect, xv, 73, 155
impoverished, 199, 256
Intelligence, 131, 199. *See also* IQ
artificial, 76, 210, 216, 218
Interactions, 38, 59, 60, 75
neural, 187
social, 10
Internal worlds, 72–77
IQ, 84, 85, 114, 140, 152, 176, 238, 239, 240
ADHD and, 255
autism and, 238–240
Raven's Progressive Matrices, 240
Wechsler Intelligence Scales, 240

Jackson, Frank, 12, 13, 14
Jackson, Michael, 80
James, Iain, 239

Kahneman, Daniel, 101–102
Kanai, Ryota, 169

Kanzi (Savage-Rumbaugh and Lewin), 197
Kawachi, Ichiro, 247
Keeling, Jeremy, 196
Kelvin, Lord, 13
Ketamine, 251
Klingberg, Torkel, 255
Knight, Bob, 175–176, 185
Knowledge, 16, 171, 269
  accumulating, 96, 107
  consciousness and, 135
  hierarchy of, 62
  strength of, 204
  structured, 75, 142
  unconscious, 95–98, 136
Kornell, Nate, 203

Lamme, Victor, 187–188
Language, 21, 46, 95, 136, 159, 194, 198, 219. *See also* Grammar; Speech
  autism and, 242
  binary, 119, 122
  body, 227
  brain and, 158
  chunking and, 151–152
  computers and, 17–20
  developmentally delayed, 241
  DNA, 45–47, 122
  learning, 151
  production, 20
Larsson, Stieg, 210
Learning, xii, xiii, 62, 125, 194, 259
  anesthesia and, 88, 91
  animal capacity for, 75, 77
  avoiding physical damage by, 66
  collective, 42
  complex, 88, 96–97
  consciousness and, 96–97
  evolution and, 63
  exploiting opportunities for, 74

forms of, 88–89
  maximizing, 126
  memory and, 21, 105, 246
  permanence of, 71
  process, 96
  rudimentary, 10, 63, 69–70, 73
  sleep required for effective, 244
  strategies, 49, 126, 136, 138, 140, 141, 141n, 143, 146, 147, 179–180, 205. *See also* Chunking
  unconscious, 89–90, 91
Leibniz, Gottfried Wilhelm von, 5
Levin, Daniel, 115
Lewin, Roger, 197
Libet, Benjamin, 103, 104, 105, 106, 125
Life. *See also* Organisms
  death barely dissociable from, 225–226
  origins of, 37–43
Limbic system, 85–86
Logic, 143, 148n
Logothetis, Nikos, 165, 167–168
Lucas, Michel, 247
Luria, Alexander, 52
Luzzatti, Claudio, 173

Mandarin, 17–20, 27
Marslen-Wilson, William, 35
Mary and knowing what it is like, 12-15
Massimini, Marcello, 215
Matata, story of, 197
Mathematics and mental arithmetic, 97, 142, 148, 176, 179–180, 182, 183
Matrix, The (film), 4
Meaning, xvi, 20, 28, 90, 134. *See also* Chinese Room argument
  consciousness and, 19

Medical Research Council (MRC), Cognition and Brain Sciences Unit, 35, 274
Medications. *See* Drugs
Meditation, 260–264
  brain changes and, 261
  consciousness and, 262, 271
  long-term shifts from, 262
  working memory and, 263
Meditations on First Philosophy (Descartes), 4
Memorization, 96, 181, 182
Memory, 59, 125, 142–143, 144, 149, 151, 188, 234. *See also* Working memory
  brain areas, 177
  chunks and, 154, 179. *See also* Chunking; Chunks
  content, 180
  effective, 244
  learning and, 246
  long-term, 136, 141–142, 150, 151, 176, 180
  loss, 84
  motor, 155
  neurons and, 121
  problems with, 221
  resources, 92
  rudimentary forms of, 63
  short-term, 137, 138, 176
  strategies with, 80. *See also* Strategies
  structured, 142, 180, 259
  superior, 52–53, 141–142, 141n. *See also* Tammet, Daniel
  tests, 170
  thought and, 177
Mental illness, xvii, 198, 220, 234, 264, 266, 267. *See also* Anxiety; Asperger's Syndrome; Attention deficit hyperactivity disorder; Autism; Bipolar disorder; Depression; Obsessive compulsive disorder; Schizophrenia
  brain abnormalities and, 254
  consciousness and, 238, 247, 263
  pandemic of, 237, 264, 268
  political neglect of, 237–238
  reinforcing by habit, 263, 270
  stress and, 261
Mental retardation, 210, 238, 239
Mental states, 1–2, 7, 12, 203
Michelson, Albert, 36
Miller, George, 32
Mind. *See also* Brain
  conscious. *See* Consciousness
  unconscious. *See* Unconsciousness
Mind-body duality, 1, 4–9
Mirror-recognition test, 130, 202–203, 210, 217
Monkeys, 203–204, 207, 208, 217
Monti, Martin, 1–2, 3, 158, 228, 229
Moore, Christopher, 181
Moral questions, consciousness and, 216–218
Morley, Edward, 36
Motivations, 70–71, 73, 176, 235, 266. *See also* Drives; Goals
Movement, 103, 227, 257
  behavior and, 69
  senses and, 68
MRI, 169, 222, 229, 231. *See also* Diffusion tensor imaging; fMRI
Multiple personality disorder, 129
Music, structures and, 95, 148, 148n
Mutations, 37, 49–53, 54–56, 61, 152n
  rates of, 58, 153, 269

Nagel, Thomas, 11, 12
Nash, John: paranoid delusions of, 256–257

Natural selection, 44, 57, 64, 84, 122.
See also Evolution
Neglect, x, 173–174
Nematode worm (*Caenorhabditis elegans*), 50–51, 70, 71, 214
Networks
brain regions, connected in, xvi, 153, 168–169, 174–176, 187–189. *See also* Prefrontal parietal network
diverse forms of, with similar characteristics, 62
fungal, 60
informational/knowledge, 16
neuronal, 41, 119–122
nodes in a, 29, 119, 190, 190n, 214–215
theories of consciousness and, 187–192, 190n, 214
Neurochemicals. *See* Neurotransmitters
Neurology, 162, 169, 174, 221, 244
Neurons, 22, 59, 60, 68, 70, 89, 186–189, 193, 211, 233, 244
activity of, 24–25, 29, 87, 90–91, 104, 105, 106, 108, 125, 126, 144, 145, 165, 194, 242, 251
artificial, 26
characteristics of, 25
communication by, 184
competition between, 118, 122–126, 134, 138, 186
computing by, 26–27
connection of, 24, 121, 190, 222
consciousness and, 184, 187, 194
cortex and, 126, 145, 167, 168, 215
in FFA, 25
flexibility in, 120–121
highways, integrity of, 230–231
influence of, 119–120
information and, 24–25, 124

landscape of, 108, 144
local, 127
machinery of, xii, 26, 106, 122, 158, 220, 265
networks of, 41, 119–122
number of, 23, 159, 160, 213–214, 215
unconscious, 86–88
waves/rhythms of, xvi, 81, 86–87, 138n, 177, 184–187
Neurophysiology, 107, 193
Neuroscience, 7, 11, 107, 119, 157–159, 264
cognitive, 130
consciousness and, x, 158, 159–162, 193
future of, 25–26
imprecise nature of, 253
Neurotransmitters, 86, 120, 122, 242, 250, 251, 252, 254, 260
changes in, 254
dysfunctional, 253, 263
New Scientist, 12n
Newton, Isaac, 239
Nikolov, Stanislav, 104, 105, 125
Nonlinguistic beings, consciousness and, 198
Nonphysical, consciousness and, 13. *See also* Mary and knowing what it is like; Mind-body duality

Obsessive compulsive disorder, 247
Octopuses, 217
brain of, 213
cognitive skills of, 212–213
Orangutans, 201, 202, 216
brain of, 212
Orbitofrontal cortex, 85, 127
Organisms, 57, 62n
complex, 63
multicellular, 59, 60, 64
viral gate-crashing of, 55

Origins of life, 37–43
Over-consciousness, 238–242
Owen, Adrian, 227, 228, 229

Pain, handling, 262, 263
Panic disorder, 247. *See also* Anxiety
Parallel architecture, 22, 23, 213. *See also* Computers; Networks
Paranoid delusions, 256–257
Paravicini, Derek, 240
Parietal cortex, 85, 139, 168, 169, 170, 173, 174, 175, 185, 189, 204. *See also* Prefrontal parietal network
Parietal lobes, complex thought and, 84
Parkinson's disease, 247
Patil, Sandeep, 252
Patterns. *See* Structure
Pavlov, Ivan, 74
Penfield, Ruth, 172, 175
Penfield, Wilder, 171, 172, 175
Peppard, Paul, 245
Perception, 132, 133, 134, 159, 163, 166n
Performance, 132, 137, 155
  improving, 141, 142, 147
(Persistent) vegetative state (PVS), 84, 222–234
  coma and, 226
  consciousness and, 231, 232
  definition of, 225
  diagnostic tests of, 230
  thalamus and, 161, 162, 225, 231, 232
  zolpidem and, 232–233
Personality, 3, 234
  changes due to brain damage, 8, 127–128, 233
  frontal lobes and, 8, 85
  orbitofrontal cortex and, 8, 85, 127–128
  problems, 107

Phillips, Patrick, 51
Philosophy, x, 4–31, 33, 107, 192
  category mistake, non-physical consciousness, 8–9
  consciousness, knowing what it is like, 11–15
  consciousness, non-physical nature, 12–15
  mind-body duality, 1, 4–9
  non-computable aspect of feelings/sensations, 15–16
  non-computable aspect of meaning, 17–21, 23, 25–28
  science versus, 4, 31, 33
Physics, 13–14, 118–119
Planck, Max, 13
Plants, information processing by, 60, 81
Posit-Science, 255
Post-traumatic stress disorder, 247, 260
Prabhakaran, Vivek, 181
Pratt, Nikki, 177
Prefrontal cortex, 85, 98, 139, 164, 168–172, 174–185, 231, 232, 251. *See also* Prefrontal parietal network
  abnormal activity in, 248, 252, 254, 260
  consciousness and, 168–171
  damage to, 172, 173
Prefrontal parietal network, 169–184, 185–191, 231, 248, 257, 258, 260, 264, 265
  ADHD and, 246
  boosting, 255
  brain-scanning, 169–171, 176–184
  chunking and, 178–184
  consciousness and, 169–171, 181, 183–184, 212
  control by, 262

damage to, 171–176
sequences and, 182
working memory and, 176
Primordial life, 37–43, 59
ideas stored in, 42, 66
Programs. *See* Computers
Propofol, 80
Protein, 6, 43, 54, 56
DNA and, 45, 88
Psychiatric conditions. *See* Mental
illness
Psychology, 93, 101, 112, 188, 193, 210
of consciousness, 109–156, 194
of unconsciousness, 88–108

Quantum mechanics, 13, 14, 187

Ranganath, Charan, 181
Rapid eye movement (REM), 246
Raven's Progressive Matrices, IQ test,
240
Recovery from brain damage, 222,
225, 231
diagnostic markers for, 223, 230
Recurrent processing model,
187–188
Rensink, Ronald, 114
Replication. *See* Reproduction
Reproduction, 36–39, 40, 53, 58, 60,
193, 266
asexual, 50–51
bacterial, 50
self-replication, 37–39
selfish, 57
sexual, 37, 50, 51
survival and, 37, 40, 54, 65, 70, 71,
193, 266
Reticular formation, 160, 161
Ritalin, 247, 250, 255
RNA, 44, 45, 55, 253
DNA and, 43, 43n, 46

Robots, 20, 28, 210
Rosenthal, David, 133
Royal Society, The, 157
Russell, Eric Frank, 31
Ryle, Gilbert, 8, 9

Sahraie, Arash, 164
Salander, Lisbeth, 210
Sarri, Margarita, 174
Savage, Cary, 181
Savage-Rumbaugh, Sue, 197
Schiavo, Michael, 223, 224, 225
Schiavo, Terri, 223–224, 225, 233
Schiff, Nicholas, 232
Schizophrenia, 98, 107, 129, 131, 240,
247, 249–252, 253, 254, 255,
256–257, 260
delusions/hallucinations and, 250
glutamate and, 252
meditation and, 263
Ritalin and, 250
treatment for, 250–251
working memory and, 249
Science, xii, 12, 14, 32, 33, 40–42, 58,
73, 98, 112, 114, 119, 134, 223
evolution and, 36
experimental failure and, 35
philosophy versus, 4, 31, 33
technology and, xv
Searle, John, 17, 19, 27–28
Seasonal affective disorder (SAD),
245, 247
Selective serotonin reuptake
inhibitors (SSRIs), 236, 254
Self-consciousness, 98, 127–134, 129,
172, 202, 219, 259. *See also*
Mirror-recognition test
awkward, 152–156
Self-recognition, 130, 202–203, 210,
217
Self-replication, 37–39

Selfish gene, 50, 57, 58
Senses, xiii, 16, 68, 144, 163, 169
Sensory features, xv, 125, 161, 163, 170, 171
Sensory input, 20, 68, 69, 82–83, 118, 186
Sequences, 141, 141n, 179. *See also* Digit Sequences; Information
  chunking, 172, 180, 183. *See also* Chunking
  digit, 150
  mathematical, 180
  patterned/structured, 142, 178, 183
  prefrontal parietal network and, 182
  unstructured, 179, 180, 183
Seth, Anil, 189n, 214–215
Sexual reproduction, 49–53, 54–56, 83, 100, 153. *See also* Reproduction
Sherashevski, Solomon, 52–53, 182
Shiv, Baba, 249
Signals, 71, 113, 124
  chemical, 10, 73, 120–121
  neuronal, *see* Neurons
  physical, 39–40
  sensory, 82–83, 118, 135
Simons, Daniel, 115
Simons, Jon, 170
Simpsons, The, 69–70
Skills, 14, 156, 206, 234, 266
  arithmetic, 182
  chunks of, 154
  cognitive, 21, 212–213
  language, 238, 239, 242
  mental, 52–53, 151, 201
  motor, 75
  sensory, 67
  social, 239, 242
  verbal, 239

Skinner, Burrhus Frederic (B. F.), 74
Sleep
  consciousness and, 245–246
  deprivation, 243, 244, 248
  mental illness and, 245–247
  problems, 242–247, 249n
Sleep apnea, 245, 246
Social decisions, consciousness and, 128
Social ranking, 49, 113
Social situations, 152, 242, 259
Son Rise Program, 239
Space Willies, The (Russell), 31
Spatial information, 115, 152, 158, 172, 175, 176–177
Speech, 3, 170. *See also* Grammar; Language
Sperling, George, 136
Stability, 42
  chaos and, 41, 44–45, 47, 154
Star Trek, 28
Stem cell therapy, 233
Stewart, Ian, 110
Stimuli, 70, 132, 144, 163, 164, 186, 263. *See also* Digit sequences; Sensory input; Spatial information
Strategies, 49, 74, 80, 143, 147, 180, 205, 264, 265. *See also* Chunking; Learning; Memory
  discounting, 138
  evolutionary, 53
Stress, 198, 262
  amygdala and, 257, 258, 260
  C. elegans sexual reproduction preference under, 51
  consciousness and, 256–260
  mutation rates in bacteria and, 49
  prefrontal cortex and, 257, 262
  mental illness and, 261

Strokes, ix, 173

Structure, 3, 61, 73, 82, 84, 97, 107, 140, 142, 149, 152, 211, 241, 263, 266. *See also* Chunking; Chunks
autism and, 250
biological value in seeking, 148
detecting, 149–151, 178, 270
exploiting, 73–74
hunger for, xiv, xv, 147–148, 155, 187, 241, 269
music and, 148
searching for, 74, 147–149, 264
self-organizing, 60
toxic thought and, 270

Subjectivity, 7, 11, 28–31, 191

Subliminal messages, 99

Superstitious beliefs, 74, 97–98, 100, 129, 250

Survival, 36, 59, 114, 193, 266
machines, 57, 58
reproduction and, 37, 40, 54, 65, 70, 71

Synesthesia, 182–183, 241, 241n

Tammet, Daniel, 182, 183, 240, 241, 242

Tang, Yi-Yuan, 263

Technology, xi, xv, 15, 21, 31, 98, 112, 219, 266, 267

Telepathy, technological equivalent of, 1–4

Temporal lobes, 84, 187

Terrace, Herbert, 205

Thalamus, 29, 84, 160–162, 165, 166, 186, 189, 194, 212–213, 225, 231, 232
consciousness and, 161, 162, 212
vegetative state and, 161, 162, 225, 231, 232

Theories of consciousness, xvi, 15, 131, 135–136, 143, 186–193, 215

Thomas, Robert, 248

Thought, 5, 16, 29, 76, 134, 155, 230, 255, 257, 267
aberrant/upsetting structures of, 263
atoms of, 118–122
brain and, 158
complex, 84, 126
conscious, 7, 49, 91, 94n, 106, 158, 199, 202, 227
decisions and, 102
experience without, 261
higher-order, 133
memory and, 177
unconscious, 49

Thought experiments, 12, 14, 17, 20, 27, 30. *See also* Philosophy

Tononi, Giulio, 189–193, 213–215

Tools, adaptive use of, 200, 201n

Transcranial magnetic stimulation (TMS), 215, 231

Turing's Nemesis, 17, 18, 20, 21, 21n, 27

Tversky, Amos, 102

Unconsciousness, xiv, xv, 7, 48, 87, 89–94, 96, 103, 106, 108, 112, 135, 150, 152, 169, 251, 267
anesthesia and, 88, 91
automatic habits and, 76, 96, 105, 147, 153–155, 177–178, 181, 184, 189, 269–270
calculation and, 94
cause and effect and, 97
chunks and, 153
consciousness and, 82, 91–94, 97, 126
control by, 98–102

Self-consciousness
decision making and, 99, 100, 101,
103, 105
information and, 92, 95
learning and, 89–90
understanding and, 90, 92
Understanding, xvii, 199, 249, 259–
260
conscious, xvi, 94n, 258
consciousness and, 90
performance and, 147
unconsciousness and, 90, 92
Unipolar depression. *See* Depression

Vegetative state. *See* Persistent
Vegetative State
Vicary, James, 99, 100
Vision, 166, 167. *See also* Visual
cortex
blindsight and, 162–164, 165, 168,
173
color, 13, 14–15, 16, 167
peripheral, 117
Visual cortex, 122, 161, 162, 163,
166n, 168, 170, 185, 188, 190
blindsight and, 162–164, 165, 168,
173
color-processing region (V4), 122,
167
consciousness and, 164–166, 188
neurons and, 167

Wallis, Terry, 233
Waroquier, Laurent, 93
Wechsler Intelligence Scales, IQ test,
240
Weiskrantz, Larry, 164
Wetware, 26, 43–45
"What Is It Like to Be a Bat?" (Nagel),
11–15

Wiener, Norbert, 110, 111, 115
Wiltshire, Stephen, 240
Wittgenstein, Ludwig, 109
Woolf, Virginia, 109
Working memory, 135, 155, 164, 175,
179, 180, 181, 185, 187, 188, 193,
204, 208, 264
ADHD and, 248, 255
analyzing/manipulating, 177, 178,
182, 186, 188, 194, 206, 208
attention and, 136, 143, 176, 262
boosting, 142, 150, 250, 255
capacity for, 141, 248, 250, 251
chunking and, 140–143
consciousness and, 144, 177, 246,
249
diminished, 172, 233, 247–252
flexible, 219
information and, 186
limits of, 137–138, 140, 143, 146,
147
meditation and, 263
objects of, 139, 140, 141
overloading, 154
prefrontal parietal network and,
176
schizophrenia and, 249
spatial, 172, 178, 250
verbal, 140, 248
World Economic Forum, 237
World Health Organization (WHO),
237

Yantis, Steven, 137

Zeidan, Fadel, 263
Zinacantecos babies/toddlers, 208
Zolpidem, 232–233